What they don't tell you about
PLANET EARTH

By Bob Fowke
Illustrations by Bob Fowke
and Andrew Mee

This book is dedicated to mad, intergalactic murderers - to keep them happy, so they leave our beautiful planet alone.

*Hodder
Children's
Books*

a division of Hodder Headline plc

Hallo, my name's *Glacier Flint*. I'm a geologist. That means I study the Earth. Come with me and we'll take a long, *cool* look at our amazing planet. Best bring a change of clothes: we'll be burning hot one minute and freezing cold the next - oh, and fasten your seatbelt. We're in for a *rocky* ride!

BAG FOR ROCK SAMPLES - AND SANDWICHES!

HAMMER FOR COLLECTING SAMPLES

Text, copyright © 1998 Bob Fowke

Illustrations, copyright © 1998 Bob Fowke & Andrew Mee

The right of Bob Fowke to be identified as the author of the work and the right of Bob Fowke & Andrew Mee to be identified as the illustrators has been asserted by them in accordance with the Copyright, Designs and Patents Act 1988.

Produced by Fowke & Co. for Hodder Children's Books

Cover courtesy of the Telegraph Colour Library.

Published by Hodder Children's Books 1998

0340 71329 1

10 9 8 7 6 5 4 3

Hodder Children's Books
a division of Hodder Headline plc
338 Euston Road
London NW1 3BH

Printed and bound by The Guernsey Press Co. Ltd., Guernsey, Channel Islands
A Catalogue record for this book is available from the British Library

CONTENTS

🦶 Watch out for the *Sign of the Foot*! Whenever you see this sign
in the book it means there are some more details at the *FOOT* of
the page. Like here.

MUM!

MOTHER EARTH - IT'S DISGUSTING!
WHY MOTHER?

Question: what have a bowl of cornflakes, a motor car and your finger nails got in common?

Answer: they all come from the Earth - maize for the cornflakes grows from the soil, basic materials for the metal and plastics of the car are dredged up from deep beneath the Earth's surface, and you yourself are mainly made up of food chemicals borrowed from the Earth and which you will give back in due course.

Almost everything that we have and everything that we are comes from the Earth. This is not a new discovery: from earliest times people have seen how the Earth gives life to plants and animals as a mother gives life to her children. Small wonder that early people came to worship the very ground they stood on as a goddess, or *Mother Earth* as she's often called.

MUM!

But what a woman! Imagine a mother 4,600 million years old and incredibly fat ...

MASS = 5,976,000,000,000,000,000,000 TONNES

AGE = 4600 MILLION YEARS

EQUATOR

CIRCUMFERENCE AT THE EQUATOR (DISTANCE ROUND THE MIDDLE) = 40 075·03 km

DEEPEST PLACE BENEATH THE OCEANS = MARIANAS TRENCH, 10900 METRES

A CASE FOR WRINKLE CREAM

Mother Earth is truly, incredibly old, in fact some scientists used to think that mountains and hills were the wrinkles in her skin caused by old-age shrinking. *Geological*  time covers such huge numbers of years that it's hard to get your mind round the lengths of time involved.

Here's one way to look at it:

Imagine Mother Earth is a middle-aged woman of 46, rather than 4,600 million, years. She's a late developer: for most of her life she's been a barren lump of rock spinning aimlessly in space. There was no life of any kind on dry land until she was *forty-two* (400 million years ago). She didn't blossom until only just *last year* (100 million years ago), that's when flowering plants

Geology is the study of the Earth. It comes from *geo*, the Ancient Greek word for earth.

first appeared on her surface. It's only *eight months* since the dinosaurs roamed, only *three days* since the first ape-men clambered down from the trees, only *ten minutes* since Jesus Christ was born in Palestine, just *a minute* since the start of the Industrial Revolution - and *less than a second* since you were born!

Fortunately, Mother Earth is expected to live till more than a hundred (perhaps another six billion years), at which time scientists think the Sun will expand and swallow her up. So there's plenty of time for her to blossom some more.

But let's not dwell on the end of it all - let's go back to the beginning ...

A *billion* is 1,000 million years.

THE BIRTH OF THE EARTH

DUST CLOUD DELIVERS LUMP OF BOILING ROCK, SHOCK!

I WANDERED LONELY AS A CLOUD ...

Once upon a time there was a lonely cloud of dust and gas (perhaps it was the remains of a star which had exploded). For millions of years our cloud drifted aimlessly through empty space with nothing to do. There were no stars nearby and nothing to cuddle up to.

Gradually the cloud began to contract. All the bits of dust and gas were pulled towards each other by the force of gravity.

Some of the dust and gas stuck together and lumps began to form. These lumps were whirled round in the cloud, gathering more gas and dust like a snowball gathering snow.

10

The biggest lump was at the centre and this lump turned into the Sun. It grew hotter as more and more dust and gas were sucked into it. Nuclear reactions started in its centre and it began to shine.

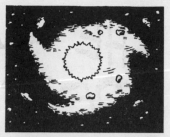

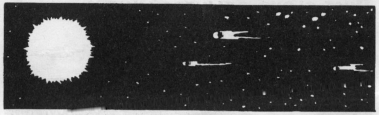

A fierce solar wind of radiation flared out from the Sun. It blew most of the light gases such as hydrogen and helium away from the innermost lumps leaving mainly heavy elements such as nickel and iron. It was these lumps which formed the inner planets, Mercury, Venus, Earth and Mars. They, and other smaller lumps called asteroids, are far heavier for their size than the outer planets.

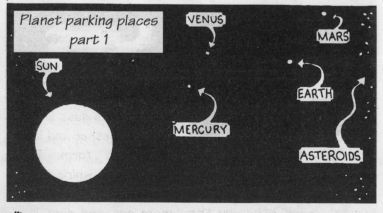

Planet parking places
part 1

VENUS
MARS
SUN
MERCURY
EARTH
ASTEROIDS

Solar means 'of the Sun'.

11

The furthest lumps from the centre became the gas giants, Jupiter, Saturn and Uranus. They're huge but they were too far from the Sun for the solar wind to blow away the lighter gases. Mostly they consist of hydrogen and helium, the two lightest gases.

SATURN

Planet parking places part 2

SUN

JUPITER

NEPTUNE

URANUS

SATURN

PLUTO- A MYSTERY PLANET- VERY SMALL AND COLD

The Sun and its planets are called the Solar System. Scientists think that the length of time between the appearance of the first lumps and when the planets reached their present size may have been no more than ten thousand years.

THE BIRTH OF THE MOON

The Moon is our nearest neighbour. It circles the Earth at an average distance of 384,400 kilometres every 27.322 days. No one knows for sure how it came into being. There are several theories:

A large chunk of the Earth broke off.

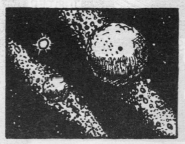

It formed from the same cloud of gas and dust as the rest of the planets and was later caught by the Earth's gravity. (This is the most popular theory at the moment.)

It formed from a different cloud of gas and dust and drifted into the Solar System.

The Inspector Calls

So there was the Earth, tumbling round the Sun as fresh as a new-born baby. A very dirty baby however, because early Earth was just a round glob of hot, liquid rock and ooze. Because it was all oozy (and still is, a lot of it) the heaviest elements soon sank down to the centre and the lightest elements rose to the surface. This is why air is on the outside of the Earth and all the rock and stuff is on the inside, with iron, nickel and other really heavy stuff in the middle - or so we think.

The problem with finding out anything about the Earth is that it's so absolutely *huge*. The deepest mine only goes down four kilometres and the deepest hole ever drilled in the Earth's surface is only fifteen kilometres, which is an amazingly long way to drill a hole but a mere pin-prick when you think that the distance from the surface to the centre of the Earth is 6,378 *kilometres* (see page 7).

14

Earth scientists have to be like clever detectives, they have to work everything out from tiny clues. As well as drilling into the land, they drill for rock samples beneath ocean beds, they measure how vibrations from earthquakes travel through the ground, they study ancient rocks and lumps of ice, they measure chemicals in the air and take samples of ancient water from deep in the oceans.

It's amazing that we know as much as we do.

Drilling a hole beneath five kilometres of water is not easy. The drill-ship has to stay directly above the drill hole whatever the weather, using sideways thrusting propellers. Nowadays they use satellite navigation systems to help the ship stay in place.

CUTTING UP MOTHER

Imagine a mad, intergalactic murderer has decided to cut up Mother Earth and have a look inside. Scientists think that the poor old thing hasn't changed very much on the inside for more than three billion years.

MOTHER EARTH

THE DIFFERENT LAYERS OF THE EARTH ARE MADE UP OF DIFFERENT MIXTURES OF ELEMENTS.

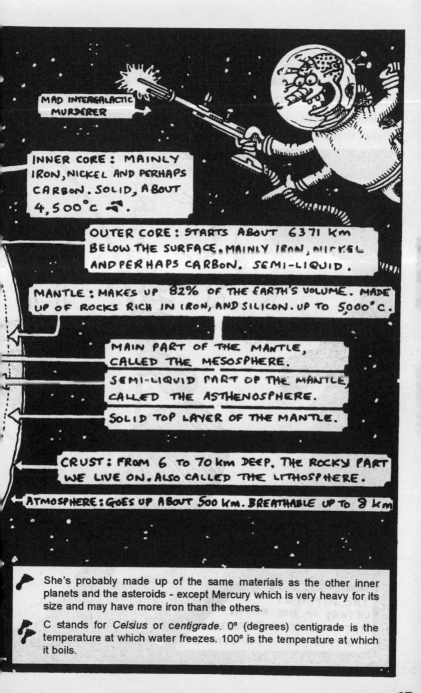

MAD INTERGALACTIC MURDERER

INNER CORE: MAINLY IRON, NICKEL AND PERHAPS CARBON. SOLID, ABOUT 4,500°C.

OUTER CORE: STARTS ABOUT 6371 km BELOW THE SURFACE. MAINLY IRON, NICKEL AND PERHAPS CARBON. SEMI-LIQUID.

MANTLE: MAKES UP 82% OF THE EARTH'S VOLUME. MADE UP OF ROCKS RICH IN IRON, AND SILICON. UP TO 5,000°C.

MAIN PART OF THE MANTLE, CALLED THE MESOSPHERE.

SEMI-LIQUID PART OF THE MANTLE, CALLED THE ASTHENOSPHERE.

SOLID TOP LAYER OF THE MANTLE.

CRUST: FROM 6 TO 70 km DEEP. THE ROCKY PART WE LIVE ON. ALSO CALLED THE LITHOSPHERE.

ATMOSPHERE: GOES UP ABOUT 500 km. BREATHABLE UP TO 8 km

She's probably made up of the same materials as the other inner planets and the asteroids - except Mercury which is very heavy for its size and may have more iron than the others.

C stands for *Celsius* or *centigrade*. 0° (degrees) centigrade is the temperature at which water freezes. 100° is the temperature at which it boils.

17

GIMME LIFE!

All life perches on the thin skin of the crust, beneath a thin layer of air, and above the huge, red-hot, oozy mass of rock and metal which makes up 99% of the Earth. Luckily for us heat only travels very slowly through rock, otherwise we'd all be burnt to cinders. In fact, heat travels so slowly from the centre of the Earth that some of the warmth now reaching us was probably created at the same time as the formation of the Solar System. It's taken 4,600 million years to reach the surface! Talk about storage heaters!

A mountain of ifs

It's a miracle there's any life at all. There is no evidence for life on the other planets, although scientists are still looking. There are so many things which could have made life on Earth impossible:

 If the crust had been even thinner.

 If Earth had been nearer the Sun.

 If Earth had been further away from the Sun.

If the Earth had been as small as the Moon, it wouldn't have had enough gravity to hold on to the air, or perhaps even the water in the oceans. Water and air would have drifted off into space.

FLINT'S QUICK FLIP

See how Glacier Flint sums up on his flip chart.

EARTH, SUN AND PLANETS ARE ALL FORMED FROM A CLOUD OF GAS AND DUST.

EARTH IS MADE UP OF CORE, MANTLE, CRUST AND ATMOSPHERE.

EARTH IS ONE OF THE INNER PLANETS.

HOW MANY TIMES HAVE YOU BEEN ROUND THE SUN?

YEARS, NIGHTS, DAYS - AND THINGS IN BETWEEN

BUT FIRST - TIME FOR A REST

It's been a hard day. All you want to do is lie down in bed. You don't want to move; you just want to lie still and go to sleep.

Fat hope! As long as you're on Earth you can never be still. Even when you're lying in bed you're moving so fast that a rocket couldn't catch you - not from a standing start.

First of all the Earth is spinning round on itself like a top, one complete spin every 23 hours, 56 minutes and 4 seconds. Each complete spin is a day. If you're at the equator the patch of Earth you're on will be moving at around 1,600 kph 👣 .

👣 kph = kilometres per hour.

Secondly, the whole Earth is circling the Sun at a speed of around 10,458 kph. Each complete circuit is a year.

Thirdly, the Earth, the Sun and the whole Solar System are travelling towards a neighbouring star, at around 69,000 kph!

Fourthly, our whole section of space is travelling around the centre of our galaxy, the Milky Way, at around 100,000 kph!

Finally, our whole galaxy is travelling away from all other galaxies at some even more incredibly immense speed!

So how old are you?

A year is one complete circuit of the Earth round the Sun, as we've already mentioned. So if you're eleven years old you've been round the Sun eleven times - simple. Each year takes 365.26 days, but the trouble is: the days may be getting longer! This may be because the Earth is growing larger, or more likely it's being slowed down by the Moon (more on that later). Either way, a year is not quite as simple a measurement of age as it seems. It's not totally clear what we mean when we say that the Earth is 4,600 million *years* old.

It's plain as daylight!

A day is the time it takes for the Earth to spin round once on itself, but that includes both day and night. Because Earth is always spinning, different parts of it face towards the Sun at different times. The darkness of night comes to you when the part of Earth on which you are standing faces away from the Sun. So at any one time, one half of the Earth is always in the shadow of night and the opposite half is in daylight.

HA! THIS'LL STOP THIS LIGHTNESS!

The Dayaks of Borneo believed that once upon a time there was nothing but daylight, until their goddess *Mang* brought darkness in a milk basket.

Of course darkness is really just absence of light, which is a form of energy. During the daytime huge amounts of energy from the Sun hit the sunny side of Earth, while the other half is dark. In fact most of the energy needed for life comes from the Sun, although some heat-energy seeps up from deep within the Earth. Sun-energy is powerful stuff: if the half of the world with Borneo on it had always faced towards the Sun and thus had always been in daylight as the Dayak legend suggests, it would have had no chance to cool down at night. Borneo would have been burned to a frazzle and the other half of the world, which would have been in permanent night, would have been a frozen wasteland. Life would probably have been impossible on both sides.

So it's a good thing that there was more than just milk in the goddess Mang's milk basket.

THE REASON FOR SEASONS

You must have noticed, unless you've got skin like a rhinoceros, that the weather gets warmer in summer and colder in winter, just as it's warmer in the daytime than at night. The reason is the same in both cases: it's all to do with the amount of Sun-energy hitting different parts of the Earth at different times.

The problem is that the Earth can't stand upright. It's tilted at an angle of 23.5 degrees from what it would be if it circled the Sun like a proper, upstanding, young planet.

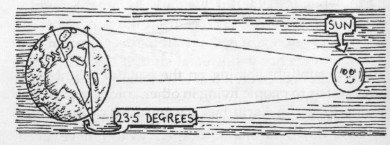

That 23.5 degrees makes all the difference - the difference between summer and winter. It means that at one point in its circuit the top half of the world catches the Sun's rays nearly full on, and later in the circuit the same thing happens to the bottom half. The north and south poles receive very little sunlight at any time - and in winter they get very cold indeed.

Around the waist of the Earth, at the equator, the tilt doesn't make much difference. This part of the world stays at roughly the same angle to the Sun at all times of the year and is always nearly broadside-on to the Sun's rays. So equatorial regions don't have winter and summer. One sun-drenched equatorial day blends into another - unless it's cloudy of course.

All that sunlight shining on the equator might seem very unfair to people living in other, colder places - but in the long run Earth has a way of balancing things out; the land keeps moving!

FLINT'S QUICK FLIP

See how Glacier Flint sums up on his flip chart.

ONE SPIN OF THE EARTH = ONE DAY.

ONE CIRCUIT OF EARTH ROUND THE SUN = ONE YEAR.

WINTER AND SUMMER ARE CAUSED BY THE ANGLE AT WHICH THE SUN'S RAYS HIT THE EARTH.

WHAT HAPPENS WHEN INDIA BUMPS ➡ INTO ASIA? ⬅

CONFUSED CONTINENTS CAN CAUSE CATASTROPHIC COLLISIONS!

BRACE YOURSELF FOR SOMETHING BIG

If you live somewhere cold, don't despair. All you need is a little patience - well, a lot of patience. The chances are that if you wait a hundred million years or so, your country may well move to the equator! Everywhere is on the move all the time. Take southern Britain: once upon a time it was where South Africa is today and no doubt it will end up somewhere warm again in the future. Meanwhile Africa is moving north and squeezing the Mediterranean Sea (which will eventually disappear) and America is drifting off to the west.

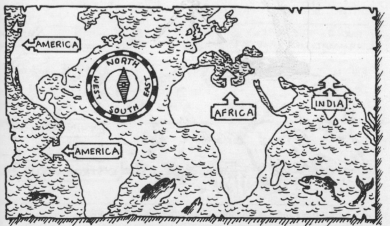

One of the fastest movers is Greece which has zoomed a hundred metres in the last three thousand years. Another speed-hog is India which is streaking northwards at seventeen centimetres per year, resulting in the Himalayas, the tallest mountains on Earth, which have been forced up where India bumps into the rest of Asia.

ONCE UPON A TIME ...

About 280 million years ago, all the continents of Earth were joined together in one huge super-continent, which scientists call *Pangaea* . Around the shores of this mighty continent washed the universal ocean, *Panthalassa*, which was the same size as all modern oceans put together. Pangaea was formed in the *Permian* period, before dinosaurs roamed and when Earth was ruled by strange animals which were neither dinosaur nor modern mammal.

Pan-gaea means 'all-land', pronounced 'pang-ee-ah'.

It's likely that the centre of Pangaea was a vast desert. The land must have been a very long way from the sea and thus from rain-bearing clouds (see page 64).

Then about 200 million years ago Pangaea started to split into two smaller super-continents *Laurasia* to the north was made up of what is now North America, Europe and most of Asia, *Gondwana* to the south was made up of South America, the Antarctic and other bits and pieces such as Australia and India. In a sense the breaking up of Pangaea is still going on. The modern continents are the flotsam and jetsam of that once-mighty super-continent.

CUTE CONTINENTS

There are seven modern continents:

Australia comes from the Latin *Terra Australis*, which means 'southern land'.

Europe comes from the ancient Assyrian word *ereb* meaning 'sunset'.

Africa comes from the Roman name of a north Africa tribe, *Afridi*, meaning 'dusty'.

Antarctica comes from Greek and means 'the opposite of *Arctic*' (the Greek name for the group of stars which includes the North Star).

North America. The Americas are named after the explorer Amerigo Vespucci (1451-1512).

South America

Asia comes from the ancient Assyrian word *asu* meaning 'sunrise'. Asia is the largest continent.

HOW EUROPE GOT ITS NAME
A Greek myth

Europa was the daughter of an Ancient Greek king. She was incredibly beautiful and Zeus, the king of the gods, fell in love with her. He turned himself into a milk-white bull and pretended to be all gentle and good (not an easy thing for a Greek god, or a bull come to that) - so gentle that Europa started stroking him and eventually climbed on his back.

Once she'd climbed on his back Zeus took off like a rocket. He carried her right across the sea to Crete, where he turned back into his god-shape and told her he loved her. But Europa was miserable and missed her home. To cheer her up the goddess Aphrodite appeared on the scene and promised that a whole new quarter of the world would be named after her. Europa then settled down in her new home and accepted Zeus.

And that's how Europe got its name.

PROVE IT!

Several hundred years ago, if you had told someone that all the continents are moving, he or she would have thought that you were stark, raving bonkers. How could something as huge and solid as a continent possibly move?

But there were certain clues. Once accurate maps were made, no one could miss the fact that Africa and America appear to fit together.

Francis Bacon, a famous English thinker, pointed this out as early as 1620, but he was way ahead of his time. The theory wasn't proved until 1915 and until as late as the 1960s most scientists were still convinced that the continents and the oceans were as fixed in place as if they had been riveted - with very large rivets.

FORSOOTH, 'TIS MOST GLARINGLY OBVIOUS UNTO ME THAT THE TWO LANDS WERE ONCE AS ONE. BUT WHEREFORE WILL NO ONE TAKE ANY NOTICE OF WHAT I'M SAYING?

The man who proved them wrong in 1915 was Alfred Wegener, the son of a German preacher, who published a book about it. Wegener looked at every bit of evidence he could find. He asked how the remains of the same types of ancient reptiles could have been found in both Africa and Brazil, he pointed out that rocks of exactly the same type could be found in America, Africa and Europe, he wondered how coal, formed in ancient, hot, wet forests, could have been found under the Antarctic ice. However, although Wegener showed that 'continental drift' happens, it took another thirty years before scientists could explain *how* it happens.

> The president of the American Philosophical Society in the 1920s was one of the scientists who continued to believe that the continents were fixed. "Utter damned rot," he said of Wegener.

SCUM ON THE POND

We now know that the continents, which are relatively light, float on top of the oozy semi-molten rock of the asthenosphere like scum on a pond. As the molten

rock slowly oozes around it drags the continents with it - and not just the continents: the entire crust of the Earth, both dry land and ocean beds, is permanently being dragged hither and thither across the surface of the Earth.

Huge 'plates' made up partly of ocean bed and partly of dry land jostle against each other. Nowadays the science of moving continents is called *plate tectonics*. *Tectonic* means 'about how things are constructed'.

There are never any gaps between the plates because as soon as a gap starts to form, new molten rock

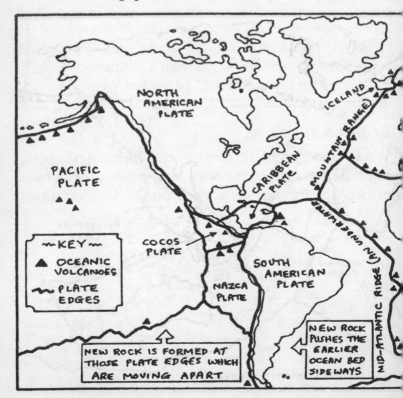

NORTH AMERICAN PLATE

ICELAND

PACIFIC PLATE

CARIBBEAN PLATE

MOUNTAIN RANGE

~KEY~
▲ OCEANIC VOLCANOES
〜 PLATE EDGES

COCOS PLATE

SOUTH AMERICAN PLATE

NAZCA PLATE

MID-ATLANTIC RIDGE (½ UNDERWATER)

NEW ROCK IS FORMED AT THOSE PLATE EDGES WHICH ARE MOVING APART

NEW ROCK PUSHES THE EARLIER OCEAN BED SIDEWAYS

bubbles up from the earth below to fill it. Iceland is a good example of land which rose relatively recently to plug such a gap.

TIME FOR BED

The whole process is powered by the ocean beds where eighty thousand kilometres of jagged, underwater mountain ranges criss-cross the planet. Some of their underwater peaks rise more than three kilometres above the surrounding ocean floor. Among these underwater mountains are many slowly erupting volcanoes which produce new crust from the hot, liquid rock below.

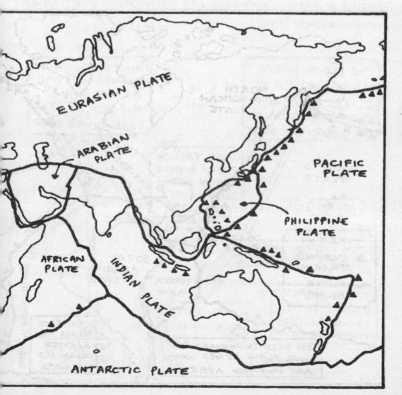

The ocean bed is heavier than the continents (even though the crust is a lot thicker under the continents at an average forty kilometres deep - under the oceans it's only an average of six to ten kilometres).

So at the edges of the oceans the heavy ocean bed plunges downwards back into the molten rock below, while the lighter continents float above. This means that the continents stay on top and keep getting older and older while the ocean beds never get the chance to get old at all - at least not by Earth standards. The oldest rocks yet discovered are on Greenland and are 3,800 million years old, but the oldest bit of ocean bed is just a 200 million-year-old baby. And when young and old meet there's often trouble ..!

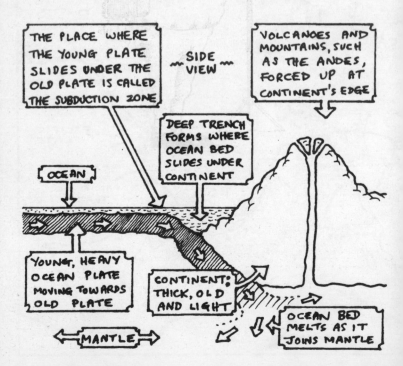

THE PLACE WHERE THE YOUNG PLATE SLIDES UNDER THE OLD PLATE IS CALLED THE SUBDUCTION ZONE

~ SIDE VIEW ~

VOLCANOES AND MOUNTAINS, SUCH AS THE ANDES, FORCED UP AT CONTINENT'S EDGE

DEEP TRENCH FORMS WHERE OCEAN BED SLIDES UNDER CONTINENT

OCEAN

YOUNG, HEAVY OCEAN PLATE MOVING TOWARDS OLD PLATE

CONTINENT: THICK, OLD AND LIGHT

OCEAN BED MELTS AS IT JOINS MANTLE

MANTLE

FLINT'S QUICK FLIP

See how Glacier Flint sums up on his flip chart.

> THERE ARE SEVEN CONTINENTS.
>
> THE CONTINENTS ARE MOVING.
>
> CONTINENTS MOVE ON TECTONIC PLATES.

DO YOU LIKE YOUR ROCK SHAKEN OR STIRRED?

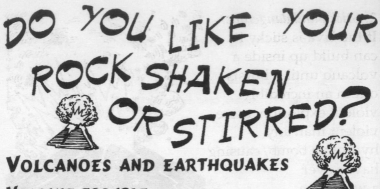

VOLCANOES AND EARTHQUAKES

VOLCANO TROUBLE

It's not very clever to live in an area where two tectonic plates meet. They're either crashing into each other (in slow motion of course), or scraping past each other or they're pulling away from each other. Either way, that's where most volcanoes and earthquakes happen. So if you live in places like Japan and California it's best to be on your guard.

Volcanoes are where liquid rock (called *magma* when it's inside the volcano and *lava* when it's outside) reaches the surface of the Earth. Volcanoes are usually mountains, but they're not all the same:

1. *Slow and steady.* Some volcanoes bubble away slowly and never cause any trouble.

2. *Sticky and dangerous.* If the lava is sticky it can build up inside a volcano until it bursts out in an incredibly violent explosion, more violent than any hydrogen bomb, causing havoc over a wide area.

ERUPTION DESCRIPTION

Scientists have different names for different types of volcanic eruption:

✦ *Icelandic*: lava flows gently from a *fissure* (a long thin gap in the crust)

✦ *Hawaiian*: lava appears gently from a vent (a central hole)

✦ *Strombolian*: lots of small eruptions

✦ *Vulcanian*: eruptions more violent than the strombolian, but less frequent

✦ *Vesuvian*: really bad news - violent eruption after long period of quiet

✦ *Krakatoan*: really, really bad news - an extra-violent explosion, may blow away half the mountain

✦ *Pelean*: violent eruption with *pyroclastic flows* (lots of lumps of hot rock or burning mud)

✦ *Plinian*: lots of lava, burning mud and anything else you care to mention

THE POSSUM PROBLEM

Volcanoes are so powerful that whole mountains can be created or destroyed within a few days, islands may appear or disappear in the sea and vast clouds of dust may blot out the Sun over half a continent. It's all very shocking - not to say very dangerous for the people who live nearby.

When volcanoes have stopped erupting they are called *extinct*. There are extinct volcanoes all over the place. For example there is a layer of volcanic rock at least three kilometres deep beneath the peaceful hills of the Lake District and North Wales in Britain. These hills are the remains of an island arc of volcanoes. But it's best to be sure a volcano is well and truly extinct before you live near one. It may just be playing possum 🐾 ...

FOOLED YA! I'M ALIVE.

WOW! A GIANT OPOSSUM! - IN WALES OF ALL PLACES.

🐾 The *opossum*, or 'possum' for short, is an American animal which pretends to be dead in order to avoid danger.

40

LETHAL LAVA

When lava flows from a volcano it may cool very quickly turning into a type of glass. This produces a weird crashing and tinkling sound as if someone is smashing up huge armfuls of china.

Pumice is a type of solidified lava with little holes in it formed by gas bubbles. (People often use pumice for softening hard skin). After the giant volcanic explosion on Krakatoa in 1883 huge floating islands of pumice were a danger to ships for miles around.

AVAST THERE ME HEARTIES!

It's hard to predict exactly when a volcano will erupt. Sometimes they rumble away for centuries producing only small eruptions or none at all. Soil from fresh lava is very fertile and often attracts large populations of farmers, so when another big eruption happens thousands of people may die. Up to fifty thousand may have died at Krakatoa and fifty-three thousand at Mimi-Yama in Java in 1793.

LAVA THIS SIDE

Lava usually flows down the sides of a volcano in red-hot streams. In Sicily there's a law forbidding anyone to try to divert a lava flow, because the diverted lava may flow on to a neighbour's property.

BEST OF BRITISH

Britain is the most interesting country in the world for geologists. It's got a bit of almost everything from every geological period. It's been repeatedly squashed between continents. In fact the highlands of Scotland were once part of America, formed when America and Europe were joined together.

This, and the fact that so many early geologists were British is the reason that several geological periods are named after British regions: Cambrian (Welsh), Ordovician and Silurian (both Ancient British tribes) and Devonian (after the county of Devon).

QUAKE TILL YOU SHAKE

The continental plates are rigid. They don't flow like water. They push or pull against each other like sumo wrestlers locked in combat; pressure builds up and then they move with a sudden jerk. Each jerk may only be a tiny fraction of a centimetre along part of a plate, but the amount of energy involved is enormous and this is what gives rise to earthquakes. The largest ever recorded earthquake, in Chile in 1960, was as powerful as a *hundred million tonne* atomic explosion.

The centre, or *focus*, of an earthquake is normally deep under ground. It's at the *epicentre*, the point on the surface directly above the focus, that the worst damage happens. The nearer the focus is to the surface the worse the damage may be.

Pressure waves appear to spread out from the epicentre like ripples on a pond, and during a major earthquake you can actually see the waves surging through solid earth. Buildings crash to the ground, fires break out, bridges buckle and fall, people are burned by the fires and crushed under falling masonry. In the most murderous earthquake ever recorded, in Shensi, China in 1556, 830,000 people died.

Seagulls, snakes, rats and dogs all seem able to detect tiny vibrations in the ground and show signs of anxiety before an earthquake. The Chinese use animals to give warning. Pheasants are the most sensitive of all.

SHAKERS AND QUAKERS

The *Richter Scale* measures the strength of earthquakes. The *Modified Mercalli Scale* is a way of listing their destructiveness ...

1. Not felt except by a few.

2. Delicate suspended objects swing. Felt by a few people at rest.

3. Definitely felt indoors. Sleepers woken, stationary cars may rock.

4. Definitely felt indoors. Sleepers woken, cars rocked, windows rattle.

5. Generally felt. Some plaster may fall, pendulum clocks stop, dishes and windows broken.

6. Felt by all. Furniture moved, plaster and chimneys damaged, objects toppled over. Many people frightened.

7. Everyone runs outside. Felt in moving cars. Structure of some buildings damaged.

8. General alarm. General damage to weak structures, not much damage to well-built structures, changes in well-water levels, sand and mud ejected, some walls and monuments fall down.

9. Panic! Total destruction of weak structures, and plenty of damage to strong structures, foundations also damaged, underground pipes broken, cracks appear in the ground.

10. Brick and frame structures destroyed, only the strongest buildings survive, foundations ruined, ground further cracked, rails bent, water slops over banks.

11. Few buildings survive, wide cracks in the earth, no underground pipes working.

12. Total destruction! Waves seen in ground, lines of sight and level distorted, objects tossed into the air.

THE ULTIMATE SURFING EXPERIENCE

When pressure waves from an earthquake hit the sea they turn into water waves - big ones known as tidal waves or *tsunamis*. Tsunamis can travel at several hundred kilometres per hour and each following wave may be up to 150 kilometres behind the one in front. They have been known to cross the entire width of the Pacific Ocean. In deep water they are quite low but when they hit shallow water near a coast their height increases and they can turn into giants up to thirty metres high. A wave like this may scoop up a ship in harbour, and carry it miles inland, perhaps dumping it on a hillside. Whole towns have disappeared in seconds. The tsunami which hit the coast of Bengal in 1873 killed 200,000 people.

FLINT'S QUICK FLIP

See how Glacier Flint sums up on his flip chart.

MOST VOLCANOES AND
EARTHQUAKES HAPPEN WHERE
TECTONIC PLATES MEET.

HOT, LIQUID ROCK IS CALLED
MAGMA OR LAVA.

THE CENTRE OF AN
EARTHQUAKE IS CALLED
THE <u>EPICENTRE</u>.

A BIT FISHY!

THERE'S OCEANS OF SPACE

DON'T FORGET YOUR UMBRELLA!

When Earth was young there were no oceans; there was just a lifeless desert dotted with thousands of volcanoes. Somehow that lifeless desert became covered in water.

One theory is that the volcanoes gushed out huge quantities of water vapour, as well as gas and lava, until at last there was so much water vapour that the air couldn't hold it any more.

Then it started to rain ...

It rained and it rained and it rained. It rained for thousands of years. Billions of litres of water fell on the desert. The water collected in the valleys and the lowlands. Ponds joined together and formed lakes, lakes grew into seas and the seas became oceans.

Another theory is that more than four billion years ago the outer layer of the mantle melted and turned into the present crust and oceans, leaving the rest of the mantle behind.

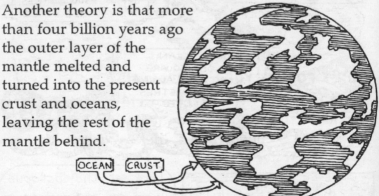

OCEAN CRUST

Either way, around 370 million square kilometres of the Earth's surface are now covered in water. That's around 1,350 million cubic kilometres of water sloshing around in the oceans, with a total weight of 1,320 billion billion tonnes.

OCEAN OCTOPUSES

Oceans move and change, just like continents. 200 million years ago Panthalassa, the universal ocean at the time of Pangea the super-continent, began to be split up by the new continents. Today there are five oceans, although they're all joined together ...

① PACIFIC OCEAN: 165,384,000 SQUARE KILOMETRES, THE LARGEST OCEAN, AVERAGE DEPTH 4,200 METRES.

② ATLANTIC OCEAN: 82,217,000 SQUARE KILOMETRES, AVERAGE DEPTH 3,600 METRES.

③ SOUTHERN OCEAN: 35,000,000 SQUARE KILOMETRES, AVERAGE DEPTH 3,730 METRES.

④ ARCTIC OCEAN: 14,056,000 SQUARE KILOMETRES, AVERAGE DEPTH 1,300 METRES, 70% COVERED IN ICE.

⑤ INDIAN OCEAN: 73,481,000 SQUARE KILOMETRES, AVERAGE DEPTH 4,000 METRES.

Look at the 'octopus numbers' above to see which ocean goes where.

NOW YOU SEA ME, NOW YOU DON'T

Oceans are huge, seas are smaller. Seas can be all on their own or they can be a part of the ocean. These seas are trying to hide from each other.

Check out the key and map below to find out which sea goes where in the real world!

1. Mediterranean
2. Caspian
3. Baltic.
4. Black
5. Red
6. North
7. Arabian
8. China
9. Caribbean

CYCLING FOR SALT

There's enough salt in the oceans to cover the whole of Europe to a depth of five kilometres, which is higher than the Alps! Most of the salt may have come from the land originally: when rain falls on land it removes minerals from the soil and rocks. The minerals are carried as salts down the rivers and out to sea. ☚ The process by which water flows between sea, land and sky is called the *water cycle*.

WATER FALLS FROM CLOUDS AS RAIN OR SNOW

WATER EVAPORATES FROM THE OCEANS, LEAVING SALT BEHIND

WATER TRAVELS THROUGH ROCK AND EARTH PICKING UP SALTS

WATER TRAVELS DOWN STREAMS AND RIVERS

WATER ENDS UP BACK IN THE OCEAN

Ostriches have special salt-secreting glands. This allows them to drink salt water.

☚ The puzzle for scientists is that the oceans don't appear to be any saltier now than they were millions of years ago - so where's all the salt gone?

SLIME TIME

Life probably started in the sea up to 2,500 million years ago. For over two billion years, until the first land plants appeared, there was no other life on Earth. Limestone and chalk are formed from the remains of underwater plants and animals, so if ever you go for a walk on limestone or chalk hills remember that what you're walking over is lots of dead bodies! Even today there are enough leftovers from living creatures floating about in the oceans to cover the whole of America under several metres of slime.

CURRENT BUN

Nearly all sea creatures live out their lives in the warm, shallow top layer of water nearest to the sunlight. Below a hundred to a thousand metres the water gets very cold and the ocean depths are an almost lifeless desert.

Below about two hundred metres the dark depths of

 We eat currants, but we swim in currents!

water circulate very slowly. It can take a thousand years for deep water to travel round the Atlantic. But, driven by wind and sunlight, the warm surface water slides swiftly over the cold, sluggish depths. It takes only a few months for water to travel along the Gulf Stream 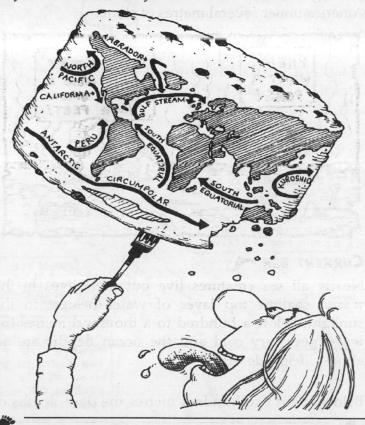, bringing warm water from the Caribbean to the north-west shores of Europe.

Major currents in Earth's oceans have flown the same way for thousands of years.

The warm Gulf Stream is why Britain never gets very cold, considering how far north it is.

Moon Themes

The important thing about the Moon is not that it looks like a cheese or, as the Japanese used to say, like a huge crystal palace belonging to thirty-six princes. The important thing is that the Moon, like the Earth, has gravity, a force which pulls at the Earth and causes the tides.

Once upon a time, the Moon may have circled far closer to the Earth than it does today, perhaps only twenty thousand kilometres away from it. At this distance it would have caused huge tides, not only in the water, but in the land as well.

Nowadays it's only water which has tides.

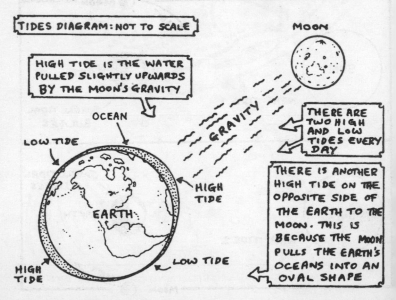

TIDES DIAGRAM: NOT TO SCALE

MOON

HIGH TIDE IS THE WATER PULLED SLIGHTLY UPWARDS BY THE MOON'S GRAVITY

GRAVITY

OCEAN

LOW TIDE

EARTH

HIGH TIDE

LOW TIDE

HIGH TIDE

THERE ARE TWO HIGH AND LOW TIDES EVERY DAY

THERE IS ANOTHER HIGH TIDE ON THE OPPOSITE SIDE OF THE EARTH TO THE MOON. THIS IS BECAUSE THE MOON PULLS THE EARTH'S OCEANS INTO AN OVAL SHAPE

The Sun also pulls at the oceans, but only weakly because it's so far away. The highest, or *spring*, tides happen when Moon, Earth and Sun are lined up and pulling together.

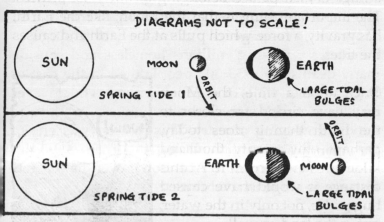

DIAGRAMS NOT TO SCALE!

SUN   MOON ⚬   ORBIT   EARTH

SPRING TIDE 1   LARGE TIDAL BULGES

SUN   EARTH   MOON ⚬   ORBIT

SPRING TIDE 2   LARGE TIDAL BULGES

Lowest, or *neap*, tides happen when the Sun, Moon and Earth are not lined up.

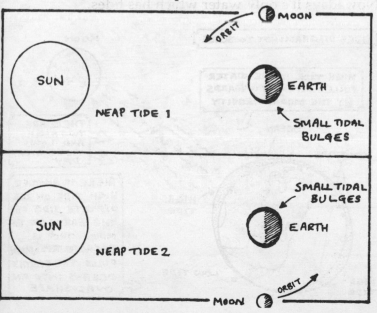

MOON ⚬   ORBIT

SUN   EARTH

NEAP TIDE 1   SMALL TIDAL BULGES

SMALL TIDAL BULGES

SUN   EARTH

NEAP TIDE 2

MOON ⚬   ORBIT

THE URGE TO SURGE

Usually it's easy to know roughly how high a tide will be, and sea walls are built to keep out the worst. But sometimes Sun, Moon and high winds all work together to produce a mega-tide or *storm surge*. In 1991 a cyclone (see page 61) caused a storm surge in Bangladesh, making millions homeless and causing many deaths. In 1953 307 people died in the east of England when walls of water, described as tidal waves by the victims, smashed through sea walls and flooded low-lying country. This disaster was one of the main reasons for building the Thames Barrier which is meant to protect London if ever the sea gets the urge to surge in the future.

FLINT'S QUICK FLIP

See how Glacier Flint sums up on his flip chart.

THERE ARE FIVE OCEANS.

WATER TRAVELS FROM OCEANS TO AIR TO THE LAND IN THE 'WATER CYCLE'.

TIDES ARE MAINLY CAUSED BY THE MOON'S GRAVITY.

IT'S RAINING SPIDERS!

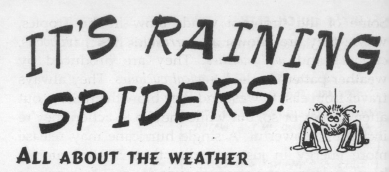

ALL ABOUT THE WEATHER

And this morning there will be showers of maggots over the southern part of the country. These will increase to a steady gale of spiders during the afternoon. But skies will clear during the night and tomorrow will be a sunny day with outbreaks of red dust. Winds will be gusty but moderate.

WHAT A BLOW!

Wind is air moving. When air doesn't move we say that it's calm. When it does we say it's windy. Wind can be strong enough to cause storm surges and weird enough to make it rain maggots as it did in Mexico in 1968, or spiders, as it did in Hungary in 1922. This sort of thing is rare but it happens because winds can carry small objects over large distances: for instance sand grains from the Sahara desert in North Africa are frequently dropped as far away as Europe, as in 1755 when red, dust-laden snow fell on the Alps.

Some of the strongest winds blow in the Tropics, where they are known as *hurricanes* , *typhoons*, *cyclones* and *willy-willies*. They are produced by weather patterns called *tropical cyclones*. They always travel from east to west and they blow themselves out after a week or so, but while they're in action they're incredibly powerful. A single hurricane may release more energy in just one day than 500,000 atomic bombs. Hurricane-strength winds can tear whole buildings to pieces and lift large lorries off the ground.

Tornadoes are smaller than tropical cyclones but wind speeds in these twisting columns of air can be even faster, reaching up to 500 kph.

Hurricanes are named after Huracan, an Arawak Indian god.

LAYERS AND LAYERS OF AIR

Air is a mixture of gases, mainly nitrogen (78.1%) and oxygen (20.9%) all mixed up into a sort of soup. The soup is thickest near the ground and that's where the weather happens.

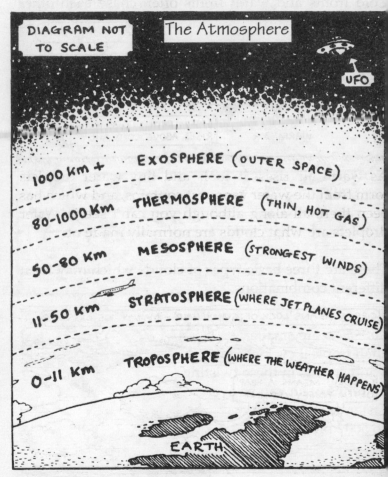

DIAGRAM NOT TO SCALE

The Atmosphere

UFO

1000 Km + EXOSPHERE (OUTER SPACE)

80-1000 Km THERMOSPHERE (THIN, HOT GAS)

50-80 Km MESOSPHERE (STRONGEST WINDS)

11-50 Km STRATOSPHERE (WHERE JET PLANES CRUISE)

0-11 Km TROPOSPHERE (WHERE THE WEATHER HAPPENS)

EARTH

A CROWD OF CLOUDS

The air in the troposphere tends to swoosh around in large lumps or *air masses*. When a warm lump meets a cold lump they don't mix; instead the hot lump tends to rise above the cold lump because hot air is thinner and lighter than cold air. If the warm lump is moving towards the cold lump this is called a *warm front*, and if the cold lump is doing the moving it's a *cold front*. Cold fronts and warm fronts often chase each other across the sky.

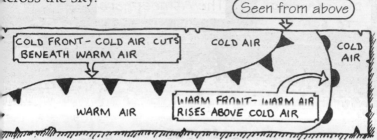

As warm air rises, it cools and tiny water droplets form from the water vapour it contains, and which has been there all along although you can't see it. Water droplets are what clouds are normally made of.

There are three basic types of clouds which may form different combinations.

Nimbus means rain-bearing. Towering storm clouds are called *cumulonimbus*, in other words, a heap of rain-bearing cloud. Most thunder storms start inside a cumulonimbus.

FEELING DEPRESSED?

Depressions are bad news for sun-lovers. When warm air gets forced upwards by a mass of cold air this means that there is 'less' air near the ground. Less air means less *air pressure* and it all adds up to a *depression*. The more cold air forces warm air to rise, the worse the depression may get.

The bad thing for sun-lovers is - the warm air cools down as it rises and clouds appear along the warm front. The clouds thicken and, hey presto, next thing you know, it's raining. Eventually, once all the warm air has been squeezed upwards the cold front of the depression, which may also be cloud-laden, catches up with the warm front. Air pressure at ground level increases and the depression is over - till next time.

WHY NOT GO TO BOGO?

Lightning is the release of an electrical charge created by water droplets in a cloud rubbing against ice crystals. If you like thunder and lightning the best place to go is Bogo in Indonesia, which has an average 332 days of thunder storms every year.

If you *don't* like lightning, why not fasten a lightning conductor to your house? This is a strip of metal which runs from the highest point of a building down to the ground. The electrical discharge of a lightning bolt is carried safely down the conductor instead of striking the building itself.

Conductors were first suggested by the American scientist Benjamin Franklin (1706 90). They were slow to be introduced into Roman Catholic countries because the people were superstitious about them, perhaps because Benjamin Franklin was a protestant. In 1776 a Catholic mob in Dijon, France, gathered to pull down a conductor.

DOWN WITH NASTY CONDUCTORS!

BOO TO PROTESTANTS!

MAIS OUI!

LIGHTNING, BY THUNDER!

Thunder is air complaining about lightning. When lightning streaks across the sky it superheats the air through which it passes and the superheated column of air expands faster than the speed of sound. Then it slows down a bit and the shock waves caused by its expansion turn into sound waves.

Thunder may be produced right down a lightning bolt. Depending on how far away it is and where it comes from it may sound like rolls, claps, peals or rumbles and often all of them at the same time. You can judge how far away you are from a thunder storm by counting the seconds between when you see a flash of lightning and when you first hear a clap of thunder. This is because light travels at roughly 299,300 kilometres a second through air, but sound only travels at 331.4 metres per second.

1...2...3...

CLAP! ROLL! THUNDER! RUMBLE! PEEL!

WOW! ONLY THREE SECONDS!
THE STORM'S ONLY ONE KILOMETRE AWAY!

SNOW, MAN

The upper air is cold. Water drops in high cloud often freeze into tiny crystals of ice. In warm weather they melt if they fall and turn into rain, but if the weather's cold they may melt and then refreeze and then they end up as snow. Snowflakes are masses of ice crystals all locked together.

The ice crystals in snowflakes tend to form around tiny 'seeds' of ice. In the sub-arctic regions of the far north, entire lakes may freeze around a single 'seed', so that the whole lake becomes one huge crystal. These lakes are thought to be the largest crystals in the world.

Jet trails are also often made up of ice crystals.

FLINT'S QUICK FLIP

See how Glacier Flint sums up on his flip chart.

MASSES OF WARM AND COOL AIR DO NOT MIX.

TROPICAL CYCLONES PRODUCE VIOLENT WINDS.

LIGHTNING IS THE RELEASE OF ELECTRICAL CHARGE CREATED WITHIN CLOUDS.

THUNDERSTORMS HAPPEN IN CUMULONIMBUS CLOUDS.

YOU ARE ENTERING THE CLIMATE ZONE!

DON'T WEAR YOUR WOOLLIES IN THE TROPICS!

MEET MR AVERAGE

There's been no rainless month recorded in Britain since 1855, so it's fair to say that Britain has a rainy climate - which is not the same thing as saying that there are no dry, sunny days in Britain. Countries don't get a different climate every time the weather changes. Climate is *average* weather:

Earth is divided into different climate zones, suitable for different types of plants and animals - such as dromedaries which store water in their bodies and live in deserts, and polar bears which live in the Arctic and have extra-thick fur to keep themselves warm.

The coldest zones tend to be in the far north and south and the hottest are near the equator, but not always. Some parts of the north or south are quite warm and some lands near the tropics are quite cold. At Tromso in Norway the lowest recorded temperature is -18°C, but in the Karakum Desert, 3,000 kilometres further south, the lowest recorded temperature is -32°C.

A world of climate zones:

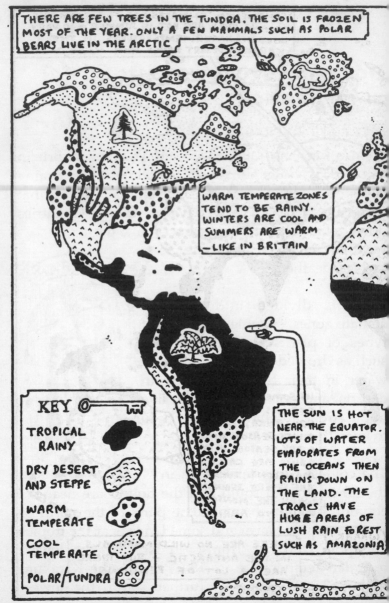

THERE ARE FEW TREES IN THE TUNDRA. THE SOIL IS FROZEN MOST OF THE YEAR. ONLY A FEW MAMMALS SUCH AS POLAR BEARS LIVE IN THE ARCTIC

WARM TEMPERATE ZONES TEND TO BE RAINY. WINTERS ARE COOL AND SUMMERS ARE WARM — LIKE IN BRITAIN

KEY

TROPICAL RAINY

DRY DESERT AND STEPPE

WARM TEMPERATE

COOL TEMPERATE

POLAR/TUNDRA

THE SUN IS HOT NEAR THE EQUATOR. LOTS OF WATER EVAPORATES FROM THE OCEANS THEN RAINS DOWN ON THE LAND. THE TROPICS HAVE HUGE AREAS OF LUSH RAIN FOREST SUCH AS AMAZONIA

'CONTINENTAL' CLIMATES ARE FAR FROM THE OCEANS. THEY ARE HOT IN SUMMER AND VERY COLD IN WINTER. TEMPERATURES IN THE KARAKUM DESERT RANGE FROM −32°C TO 54°C

IN SOME TROPICAL REGIONS SUCH AS SRI LANKA, THERE IS A WET SEASON AND A DRY SEASON. HEAVY RAINS ARE CALLED MONSOONS. THERE ARE SEVERE DROUGHTS WHEN THE MONSOON FAILS TO ARRIVE

WAAAK!

THERE ARE NO WILD MAMMALS IN THE ANTARCTIC — BUT THERE ARE A LOT OF PENGUINS!

DESERT DUST DOWN

Deserts are the driest regions on Earth and the most unfriendly to life. And they're getting bigger: the Sahara increases by an area the size of Wales every year. It's already as big as the USA.

There are six hot deserts and four colder deserts. The six hot deserts in this picture are beating up the four cold deserts, because they think the cold deserts are trying to cool them down.

— Where they are in the real world!

WADI YOU THINK OF THIS?

Some dire desert dangers

Desert winds blow sand into *dunes*. These are a bit like waves in water, but solid and they can be as big as small hills. The winds often create blinding dust storms when it's impossible to see even a few metres ahead. Fortunately desert sand feels less gritty than sand from the seashore. This is because the sand gets rubbed smooth and round by being blown against other sand grains in the dry climate.

You might think that living in a hot desert you would at least stay warm, but that's not always the case. Deserts have the world's biggest differences between day and night time temperatures. Even in the Sahara the nights can be so cold that they're frosty. This is because there's no cloud cover to trap the daytime heat.

Most deserts are criss-crossed by dry river beds, known as *wadis*. The land is too dry to soak up water quickly. So on the rare occasions when it rains the water runs off the land and may surge down the wadis in *flash floods* which sweep all before them without warning. Experienced travellers never camp in wadis.

FLINT'S QUICK FLIP

See how Glacier Flint sums up on his flip chart.

CLIMATE IS AVERAGE WEATHER.

DIFFERENT CLIMATE ZONES ARE SUITABLE FOR DIFFERENT TYPES OF PLANTS AND ANIMALS.

THERE ARE TEN DESERTS.

WADIS CARRY WATER FROM FLASH FLOODS IN DESERTS.

MOUNTAINS OF MOUNTAINS

GUESS WHAT - ALL ABOUT MOUNTAINS!

BABY!

The *Rocky Mountains* (Rockies) which run from the USA into Canada are babies. Fifteen million years ago they didn't exist, but like all babies they're growing fast - roughly 2.5 centimetres per year. The youngest mountains are always the tallest because wind and weather have not had much time to wear them away.

On the other hand the mountains in the centre of Canada are very old - so old that they're not mountains any more. They've been worn down to the roots. Yes, mountains have roots just like teeth - Earth's crust is up to fifty kilometres deeper under mountains than it is under the rest of the continents. The weight of mountains pushes the crust deeper into the molten rock below.

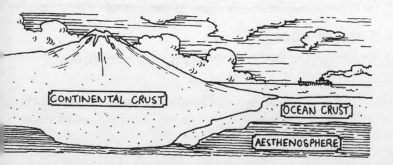

75

MOUNTAIN MOLARS

Imagine mountains really are teeth - false teeth!
Can you see which of these greedy people has the
largest dentures? (Answers upside down)

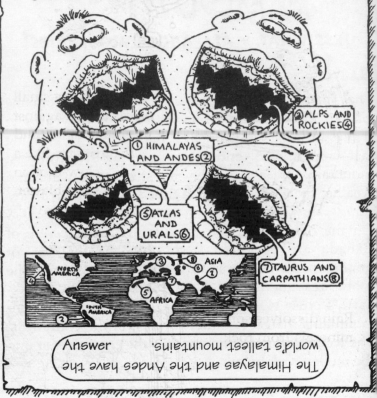

② ALPS AND ROCKIES ④

① HIMALAYAS AND ANDES ②

⑤ ATLAS AND URALS ⑥

⑦ TAURUS AND CARPATHIANS ⑧

Answer
The Himalayas and the Andes have the world's tallest mountains.

SOFT AS ROCK

Many things *erode*, or wear away, mountains:

The roots of plants penetrate loose outer flakes of rock.

When ice feezes it expands. If it freezes in a crack inside rock it may cause chips of rock to break off.

Winds blow small pieces of grit against mountain sides, and the grit acts like a slow-motion sand-blaster.

Rain dissolves some minerals from rock.

ICE PAGES

Ice has probably done more to wear away mountains than anything else. Not just puny slithers of ice, but massive great rivers of the stuff which eat away whole mountains.

Every few hundred million years the Earth catches cold. In other words, it suffers a series of *ice ages*. The last one reached its peak only 18,000 years ago. At that time nearly a third of dry land was frozen over, Britain was covered in ice as far south as the Bristol Channel, and even North Africa was a lot colder.

HERDS OF WOOLLY MAMMOTHS ONCE ROAMED BRITAIN. THEY MAY HAVE LOOKED FOR FOOD BY USING THEIR TUSKS TO SCRAPE AWAY THE ICE-AGE SNOW. IT'S THOUGHT THAT MAMMOTHS WERE MADE EXTINCT BY STONE-AGE PEOPLE WHO OVERHUNTED THEM!

At present the ice is still retreating. Scandinavia is growing taller because it's much lighter than it used to be as a result of the ice which was on it melting. But don't get too hopeful: we could be due for another ice age at any time.

It's *glaciers* which do the damage. Glaciers are slow-moving rivers of ice which flow at speeds from a few centimetres a year to about two metres per day, picking up loose rocks as they go. The loose rocks crushed beneath glaciers act like heavy-duty sandpaper, perfect for scraping away mountains. The more rock the glaciers scrape away, the more rock they have for scraping. They can't lose.

GLACIERS

GLACIER SNOUT

ROCK AND CLAY CARRIED BY A GLACIER IS CALLED MORAINE

WHEN THE CLIMATE BECOMES WARMER, THE GLACIER WILL RETREAT AND THE MORAINE IS LEFT BEHIND

Take a look at a mountain next time you get the chance. Very likely it will have been scraped out into huge curved shapes called *cirques* . Often glaciers have been at work back to back leaving cirques on every side with only narrow ridges in between, called *aretes*. Sometimes the glaciers have done so much scraping that only a steep horn of rock is left, such as the *Matterhorn* in the Alps.

CIRQUE

CIRQUE

ARETE

Also called *corries* in Scotland and *cwms* in Wales.

FLINT'S QUICK FLIP

See how Glacier Flint sums up on his flip chart.

THE TALLEST MOUNTAINS ARE THE YOUNGEST.

MOUNTAINS HAVE ROOTS. THE EARTH'S CRUST IS DEEPEST UNDER MOUNTAINS.

GLACIERS ARE SLOW-MOVING RIVERS OF ICE.

GLACIERS SCRAPE AWAY THE SIDES OF MOUNTAINS.

THE RAIN DRAIN

RIVERS RUN THEIR COURSE

RUNNING WATER

The ice ages have affected more than mountains. Ice has made great rivers change course. Before the last ice age the River Severn flowed north into the Dee estuary south of Liverpool, England. Then ice blocked its course like a huge dam. The water built up into a large lake and eventually broke through the Ironbridge Gorge to the south. It now flows south to Bristol. Likewise, until ice changed its course the River Thames flowed to the north east avoiding the area that would later become London.

But rivers are not always a walkover. They don't always turn around and go the other way at the first obstacle. Rivers have done even more to change the land than glaciers, carving out huge valleys and leaving flood plains behind them. Take the River

Indus and the Himalayas. The Himalayas are big, baby mountains like the Rockies, but the Indus which flows *through them* is much older. It cuts through the rock faster than the mountains can grow, so that it keeps on flowing south down many deep canyons - in spite of the Himalayas.

DRAINS AGAIN

Every river has its own *drainage basin*, which is not a large sink with drain water in it. A drainage basin is the area of land from which rain water drains into a river . Because water flows downhill there's always a boundary ridge or *watershed* between one drainage basin and the next.

WATERSHED | A DRAINAGE BASIN | WATERSHED
TRIBUTARIES

From distant parts of a *drainage basin* little streams drain into larger streams. These larger streams join together into small rivers, which drain into larger rivers, and eventually all the water is collected into one large river which drains into the sea. The smaller rivers which drain into larger rivers are called its *tributaries*.

In some places the water collects underground. Huge quantities of water up to 25,000 years old have collected beneath the Sahara desert and there it stays, locked in the rock, unless people pump it out.

SNAKES AND RIVERS

Snakes are long and wriggly, but rivers are a lot longer and wrigglier. These five famous rivers have been joined together by canals, but blocked by snakes. See if you can get from the watershed to the sea - without crossing any snakes!

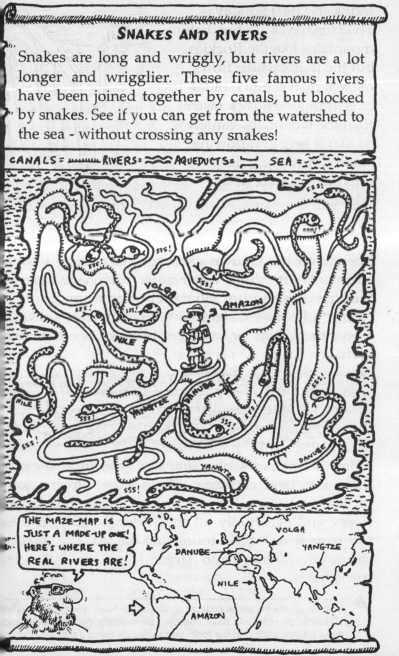

ROCK TILL YOU DROP

Water flows fastest when it's falling most steeply down a hill. The faster it flows the rougher it tends to be. A rough mountain river will pick up pieces of rock from a river bank and roll and carry them tumbling downstream. But once the land starts to flatten out, the water calms down and starts to drop whatever it's carrying. The biggest rocks get dropped first and the smallest get dropped as *silt* closest to the sea. This is why mountain streams are cluttered with large boulders but large lowland rivers tend to have fine clay banks.

Lowland rivers slow down so much they may start to *meander*. They weave around like snakes looking for breakfast, leaving silt and mud behind them.

They often overflow on to a *flood plain*. This is an area of land which has been flattened out by all the silt which is left on it. Flood plains make very rich farm land, and this has led to some of the world's worst disasters. Lots of people tend to live on them because the land is so rich and well watered - but they're still *flood* plains. In 1931 the *Huang Ho* in China flooded, killing several millions. This was possibly the worst flood disaster of all time.

See how Glacier Flint sums up on his flip chart.

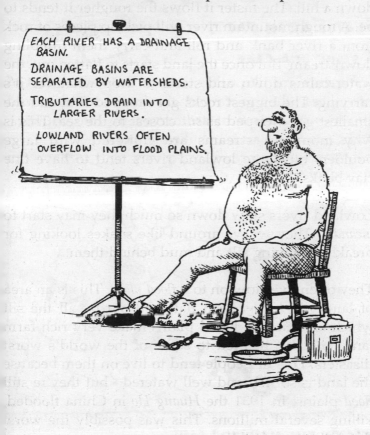

EACH RIVER HAS A DRAINAGE BASIN.

DRAINAGE BASINS ARE SEPARATED BY WATERSHEDS.

TRIBUTARIES DRAIN INTO LARGER RIVERS.

LOWLAND RIVERS OFTEN OVERFLOW INTO FLOOD PLAINS.

MAPPING IT OUT

FLAT EARTH FACTS

A STAKE IN A CAKE

Continents are like cakes, except they're not carved into slices, they're carved up into countries, and countries in their turn are carved up into counties, states, provinces and suchlike divisions.

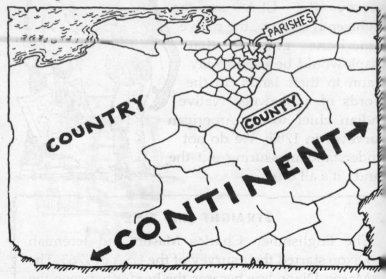

IT'S ON THE MAP, YOU SAP!

Countries are not natural like continents or mountains: they're man-made, political divisions of the land, and maps of the world are very political things. Countries go to war over maps, or rather over the borders described by them. This is why although borders are man-made they often follow natural boundaries which are easy to defend, such as mountains and rivers - at least they used to ...

Modern maps are not always a blessing. Politicians have been able to draw borders in straight lines, ignoring all the natural twists and turns of rivers and mountains which may have formed natural boundaries between groups of people.

And without maps it's very hard to prove who owns what piece of land. When the Americans started to survey the USA in the eighteenth century, the native Americans guessed that maps would be used to lay claim to their lands. In the words of a Shawnee Native Indian chief to an American surveyor in 1785: 'We do not understand measuring out the land - it's all ours.'

IT SAYS SO ON THE MAP- SO IT'S OURS, SEE?

STRAIGHT LINE TIME

The Englishmen Charles Mason and Jeremiah Dixon started their survey of the USA in 1763. The Mason-Dixon line became the border between the northern 'free' and southern 'slave' American states from about 1820, and was the line which divided the two sides at the start of the American Civil War.

MASON-DIXON LINE

It's central

The earliest map of the world was carved on to a tiny clay tablet in ancient Babylonia. It shows the Earth as a flat disc with Babylon at the centre.

Before modern maps, most people thought that their own country was at the centre of the world. The Ancient Greeks believed that Mother Earth had a tummy button at Delphi in Greece, the Japanese thought the centre was Mount Fuji, and as late as the sixteenth century an Italian priest complained that the Chinese always drew the Universe as being made up of China with just a few islands in the surrounding ocean for all the other countries!

CONQUEST!

You aim for world domination. The seven continents shown on this map are yours - if you can answer the following questions.

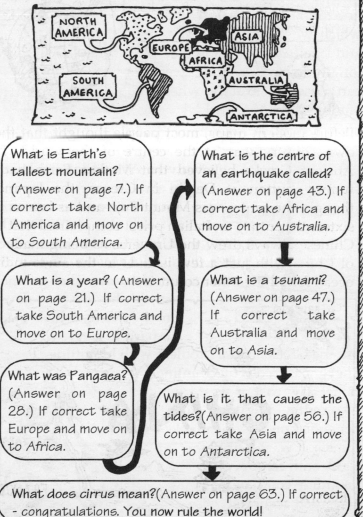

What is Earth's tallest mountain? (Answer on page 7.) If correct take North America and move on to South America.

What is the centre of an earthquake called? (Answer on page 43.) If correct take Africa and move on to Australia.

What is a year? (Answer on page 21.) If correct take South America and move on to Europe.

What is a tsunami? (Answer on page 47.) If correct take Australia and move on to Asia.

What was Pangaea? (Answer on page 28.) If correct take Europe and move on to Africa.

What is it that causes the tides? (Answer on page 56.) If correct take Asia and move on to Antarctica.

What does cirrus mean? (Answer on page 63.) If correct - congratulations. You now rule the world!

A GNOME ON A GNOMON

It was probably the Ancient Greeks who first worked out that the Earth is round. And it was an Ancient Greek, Hipparchus (died 125 BC), who had the brilliant idea of dividing the world into horizontal and vertical lines, which we now call *latitude* and *longitude*. By giving the longitude and latitude we can describe exactly where anything is in the world.

It was easy to measure latitude. At midday the length of a shadow cast by the Sun gets longer as you move further away from the equator. The Ancient Greeks used a stick called a *gnomon* to make a shadow. The gnomon could be up to eleven metres long, which made it hard to carry.

GNOMON

SMALL GREEK GNOME

SHADOW

But the Greeks had a problem with longitude. They measured it by counting camel days east or west from Alexandria in Egypt - which depended on undependable camels. It wasn't until the eighteenth century that longitude could be measured accurately.

Pronounced *no-mon*.

Maps may cause wars, but accurate maps can also *save* lives. More than four hundred pilots per year used to die in Bolivia by crashing into incorrectly mapped mountains or trying to land by incorrectly mapped villages. Now that Bolivia is accurately mapped there are very few crashes.

TROUBLE WITH TRIANGLES

If you know the length of one side of a triangle and the angles of two of the corners it's possible to work out the lengths of the other two sides using *triangulation*. By standing on one hill and looking at two other hills you can work out the exact distance to one of them without moving, provided you have already measured the exact distance to the other one.

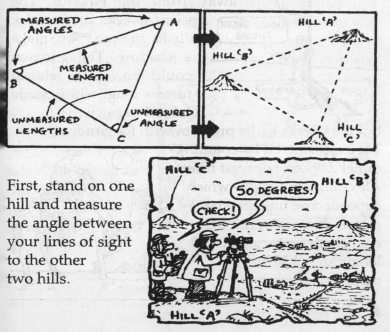

First, stand on one hill and measure the angle between your lines of sight to the other two hills.

Next, walk to the next hill, measuring the distance between them.

Then, measure the angle between your lines of sight to the two other hills.

You can now calculate the distance from where you are to the hill you haven't yet visited - and you can use your newly calculated distance as the 'measured' side of a new triangle - and so on until you've covered an entire country in triangles and you know the exact position of each point of each one. And once you've

done that you can easily measure the exact distance of anything from the nearest *trig point* ,
as the points of the triangles are known - and you can make a very accurate map.

Trig points are often marked with lumps of concrete or stone. The first British ones were positioned by the Ordnance Survey in 1791. You can find trig points on the tops of many hills in Britain.

France was the first country to be accurately surveyed using triangulation. When the survey was completed in May 1682 King Louis XIV was horrified to find that what he had thought was his coastline had 'shrunk' by up to 160 km in places. He said it had cost his kingdom more than losing a war!

GET THE SCALES OUT!

Maps done to size would be very awkward to use. Imagine having to look at a map of Australia which was the same size as the actual country - you might as well look at Australia itself!

Map makers get round this problem by drawing their maps 'to scale'. The principle is very simple: you simply say 'let one centimetre (or whatever measure you choose) on the map be worth one kilometre (or whatever measure you prefer) in the real world'. Or, putting it another way, you can say that a map is 5,000 to 1 meaning that one unit on the map (whether it's a centimetre or a kilometre doesn't matter) is worth

5,000 in the real world. Next time you go trekking make sure you know the scale of your map - otherwise you could be in for a very long walk home (or a very short one)!

NOT AS FLAT AS ALL THAT!

Maps describe the rounded surface of the Earth, usually on a flat sheet of paper. But they can also show hills and valleys. Some maps use shading to describe them, but others use *contour lines*. These are lines which show height above sea-level. If they're close together, this shows a steep hillside.

If the contour lines are far apart this means that there's only a gentle slope.

And if there are no contour lines, this means the Earth really is flat!

FLINT'S QUICK FLIP

See how Glacier Flint sums up on his flip chart.

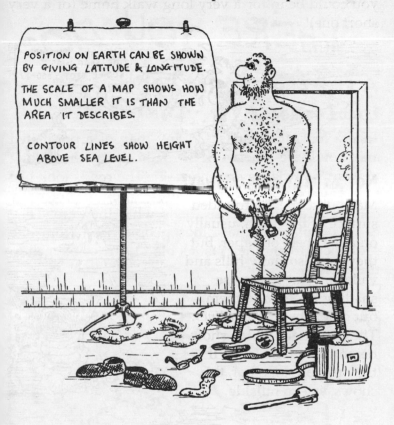

POSITION ON EARTH CAN BE SHOWN BY GIVING LATITUDE & LONGITUDE.

THE SCALE OF A MAP SHOWS HOW MUCH SMALLER IT IS THAN THE AREA IT DESCRIBES.

CONTOUR LINES SHOW HEIGHT ABOVE SEA LEVEL.

WHY DO YOU LIVE WHERE YOU DO?

A PITY ABOUT CITIES

LOADS OF ROADS

Take a look at an atlas. On any map of a populated area you'll see a network of roads and paths linking towns , villages and cities . The roads look like blood vessels, especially since they're normally coloured red. In a sense that's what they are. They carry people and the things people need from one settlement to another, just as blood carries oxygen to the muscles.

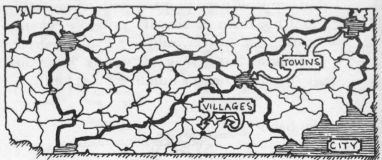

Look at the map again and you'll see that there are lots of villages, fewer towns and even fewer cities: the bigger the fewer - and also the further apart. It stands to reason really, but have you ever thought why?

When does a village become a town? Different cultures have different ideas. The Japanese say when it has a population of 30,000 people, the Danes say when it has 250 people!

The word city used to mean a town with a cathedral, but now it tends to mean any very large town.

WHO GOES WHERE

Once upon a time everybody walked to market, or they were carried there in a horse and cart. That's why there tends to be a small market town every ten to twenty miles in populated regions, especially in Britain. The market towns had to be near enough to the villages and farms so that people could go there, do their business and get home again, all in the same day.

Some things such as courts of law and bigger shops weren't available in small towns. For those things people were prepared to travel further to a big town or city. It's not so different nowadays: you would walk a hundred metres to buy a bag of sweets, but you wouldn't drive a hundred miles for it.

On the other hand you might well travel a hundred miles to buy a special new bicycle and even further to see a football match or a show - or to go to court if you had been accused of murder.

ALL CHANGE!

Towns and villages come and go. There are thousands of abandoned villages in England, just mounds of grass where once there were houses and shops. Once a village has lost its purpose, it will die if it can't find another one. Once upon a time villages were homes for people who worked on the land, but in industrialised countries that's changing.

Village voices

A. Why buy food in the village shop when you can jump in your car and fill up the boot at the nearest supermarket?

B. Who needs lots of farm workers when machines can do the job?

C. Why live in a crowded town when you can buy a nice village house no longer needed for a farm worker and drive to your work in town?

D. Why not buy a holiday home in a pretty village?

A + B + C + D = a commuter village for the wealthy.

TOWN TIPPLE

In some advanced countries there is a movement of city people to the country, but in most of the world the really big change is happening the other way round.

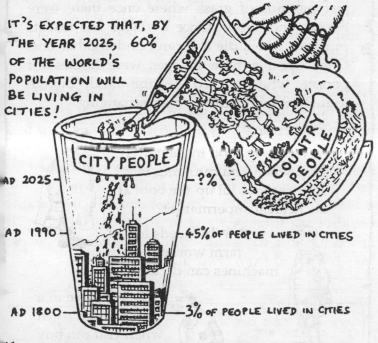

IT'S EXPECTED THAT, BY THE YEAR 2025, 60% OF THE WORLD'S POPULATION WILL BE LIVING IN CITIES!

CITY PEOPLE

COUNTRY PEOPLE

AD 2025 — ?%

AD 1990 — 45% OF PEOPLE LIVED IN CITIES

AD 1800 — 3% OF PEOPLE LIVED IN CITIES

NOT SO PRETTY CITIES

Cities are growing bigger and bigger. In 1900 only London and Paris had populations of more than one million. Today there are nearly three hundred cities of more than a million in the world and several have populations of more than ten million. That's a lot of people.

Cities have become so huge that they create their own climates. They have less sunshine and more cloud and rain than surrounding areas, but at the same time they're warmer. Cities are like great big heaters. Traffic creates heat, so do the people. This man-made heat as well as heat from the Sun is stored in the city's walls, roofs and roads during the day then released slowly at night. Night time temperatures can be 4°C warmer than in the countryside. So if you want to be warm and wet (well, warmer and wetter) go and live in a city.

There's a more unpleasant difference between many cities and smaller towns and villages: smog. It's not new. When coal was the most commonly used fuel there used to be terrible smogs in London. *Peasoupers*, as they were called, were thick yellow fogs caused by dirt from coal fires. The last was in 1952 when the fog was so thick that if it got into a house, the inhabitants couldn't see from one end of a room to the other. It lasted for four days and up to four thousand people may have died from chest infections.

Queen Elizabeth II was *crowned* in 1953! (Although she became Queen when her father died in February 1952.)

Coal has gone out of fashion, but modern cities still produce tonnes of dust and dirt from exhaust fumes and chimneys of all kinds. A great dome of dusty air hangs above most cities unless there's a strong wind to blow it away. In Tokyo, people often wear smog masks and there are coin-operated oxygen machines for people who are about to keel over from traffic fumes. Which is not to say that Tokyo is dirtier than other cities: it's just better prepared.

Well, that's the run down on the bad things about cities. But let's not forget all the good things – which are why so many people live in them after all: things such as theatres, cinemas and jobs, to name but three.

FLINT'S QUICK FLIP

See how Glacier Flint sums up on his flip chart.

THERE ARE LOTS OF VILLAGES,
FEWER TOWNS AND EVEN
FEWER CITIES.

ABOUT HALF THE WORLD'S
POPULATION NOW LIVES IN CITIES

SMOG IN CITIES IS CAUSED BY
EXHAUST FUMES AMONG OTHER
THINGS.

TOO MANY  PEOPLE?

PEOPLE POSE PRETTY PERILOUS PROBLEMS FOR PLANET EARTH, PROBABLY

HALLO HUMANS – THE STORY SO FAR

100,000 BP (Before Present). Nobody knows for sure when modern humans first appeared on Earth. They probably started in Africa and first shook their shaggy locks about a hundred thousand years ago - a few hours ago by Mother Earth standards.

40,000 BP. For perhaps sixty thousand years they lived quietly in Africa, content to hunt hippo and dine on rhino. Then around forty thousand years ago they got itchy feet - no one knows why. Waves of humans spread out from Africa and colonised every part of the world except Antarctica. They, or rather we, turned out to be a very tough species. We even survived in frozen northern Europe during the height of the last ice age around eighteen thousand years ago, the coldest period people have ever known.

For another thirty thousand years humans continued to be simple *hunter-gatherers*. Small family groups roamed about gathering food from the wild animals and plants around them, but growing nothing of their own.

10,000 BP. About ten thousand years ago people started to plant their own crops. Humans, being much smarter than all the other animals, soon realised that this was a much more efficient way to get food than hunting and gathering. The result of this was the *Neolithic Revolution*, as the start of farming is now called. From now on there was no need for humans to wander from place to place, in fact it was easier to set up a village - or a town. More food could always be grown to feed increasing numbers of people.

Civilisation was on its way.

Boom!

When humans first spread out from Africa there were perhaps just two million people in all the world. When farming first started there were probably only five million. Even by the time of Christ, the total world population was no more than 200 million. This is peanuts by modern standards.

It was only after the *Industrial Revolution* of the eighteenth and nineteenth centuries that world population really *boomed:* during the 1900s world population topped a billion and it's kept growing ever since. In 1987 the United Nations estimated that the *five billionth* person had been born. Population is now growing at a rate of *150 people per minute*, or *96 million per year*. The world is crawling with people like ants on an antheap - no disrespect to ants!

Rubbish!

People make pollution. It seems that unless we're very careful almost everything we do causes pollution.

☠️ Beside every big city there's a huge rubbish tip. Greater New York for instance produces more rubbish than the sediment from all rivers from Maine to North Carolina. Rubbish tips leak methane gas and other chemicals. If there's no rubbish tip beside a city then the rubbish is probably being burned, which produces other polluting gases.

☠️ Petrol and diesel-driven vehicles and coal-fired power stations give off sulphur dioxide and nitrogen oxide. These gases can be carried through the air for long distances. Eventually they combine with water vapour to produce dilute sulphuric and nitric acid and fall as *acid rain*, killing trees and fish. Many Scandinavian rivers have no fish left, and it's likely that there won't be many trees at all left in Germany a hundred years from now, if nothing is done.

☠ The burning of wood, coal, gas and petrol produces carbon dioxide. Carbon dioxide in the air traps the Sun's heat and stops it escaping into space. There is now 15% more carbon dioxide in the air than there was in 1900 and the world is heating up like a greenhouse. If it heats up too much the ice at the poles may melt, causing the sea level to rise and drowning many coastal cities. Climates will change and there may be serious storms and floods.

☠ Chlorofluorocarbons (CFCs) in fridges and some aerosol sprays destroy ozone high in the atmosphere. This ozone protects the surface of Earth from damaging ultraviolet rays from the Sun. Too much ultraviolet causes skin cancer in people, and if there was no ozone at all life on the Earth's surface would be impossible. Nowadays governments are trying to control the use of chlorofluorocarbons.

SUICIDE IS NOT A SENSIBLE SOLUTION

If there were less people in the world, there might be less pollution. However, if you're a person and you're reading this book, don't go out and kill yourself. One in five billion won't make much difference. It would be better to hang around for a while and do something positive to make the world a better place!

FLINT'S QUICK FLIP

See how Glacier Flint sums up on his flip chart.

WORLD POPULATION IS GROWING AT THE RATE OF 150 PEOPLE PER MINUTE.

ACID RAIN KILLS TREES AND FISH.

EXCESS CARBON DIOXIDE MAY BE HEATING THE WORLD LIKE A GREENHOUSE.

CFCs DESTROY OZONE.

DOWN ON THE FARM

ALL ABOUT LIFE - AND HOW TO MAKE THE MOST OF IT

LIFE CHANCES

After the volcanic eruption of 1883 (see page 41) the island of Krakatoa ended up a barren lump of ash and rock which had shrunk to a third of its former size. No plants or animals were left alive. But within just three years twenty-six species  of animal had reappeared. Fifty years later there were well over a thousand different species of plant, animal and insect on the island, buzzing, clucking and grunting, and doing all the other things that living things do.

There is a maximum amount of life that any piece of land can support given its climate. Left to itself Krakatoa will reach a new *climax* of life. So would the rest of the Earth - if it had the chance. But most of the Earth *doesn't* reach a climax of life. For example, huge areas of the globe should be covered in rich, lush forest right now - but something's stopping that happening.

A *species* is a type of animal or plant, such as lion or lettuce.

109

FAT FARMERS AND THIN FARMERS

What's stopping it happening is mostly *farmers*. Farmers try to limit the number of species of plants and animals on their land so they can grow food for us to eat. It's understandable - who wants outsize oaks in their oat fields, wild wildebeests in their wheat fields or big cats in with the cattle?

Much of the landscape that we see today has been farmed for centuries. It looks completely different to

KEY

NOMADIC HUNTERS	1
NOMADIC HERDERS	2
SHIFTING CULTIVATION	3
INTENSIVE SUBSISTENCE	4
COMMERCIAL PLANTATIONS	5
LIVESTOCK RANCHING	6
COMMERCIAL CEREALS	7
MIXED FARMING	8
'MEDITERRANEAN'	9
IRRIGATED FARMING	10
UNSUITABLE FOR FARMING	11
MANAGED FORESTS	12

'MEDITERRANEAN' FARMING IS MAINLY WINE, OLIVES AND CITRUS FRUITS

'SLASH AND BURN' FARMING IS CAUSING DEFORESTATION IN AMAZONIA

how it would have done if it had been left alone.

It's not just farmers who change things. Land which hasn't been farmed may well have been worked on by someone else: hillsides have been dug up by miners, sea shores protected with sea walls and jetties, more than half of Earth's rivers dammed. In fact landscapes which seem to be completely natural may be nothing of the sort. The Norfolk Broads, a system of inland waterways, were created by medieval peat diggers.

RICE IS A MAJOR 'IRRIGATED FARMING' CROP IN CHINA, INDIA AND SOUTH·EAST ASIA

'PLANTATIONS' ARE MAINLY CROPS OF TEA, COFFEE etc.

TAKE A LOOK AT PAGES 70 to 71 — AND SEE HOW CLIMATE AFFECTS FARMING!

Fancy some salt on your wheat?

Some of the earliest farmland on Earth is now desert. The early civilisations of Mesopotamia (in modern Iraq) and the Indus valley (in modern Pakistan) kept their farms going with *irrigation* systems because not enough rain fell in those parts of the world. Water was channelled from rivers and dribbled on to the fields to produce rich crops.

When rain water drains directly from rivers into the sea, the salt collected on its journey is dumped in the sea (see page 53). But unfortunately, if the water is left to dry in the fields then the salt is left behind. As years go by the salt builds up in a process called *salination* - until nothing much will grow any more.

IS THIS MEANT TO BE LUNCH?

This will probably happen to parts of America before long. In the last thirty years, except for a few floods, no water from the Colorado River has reached the sea. It's either dammed to produce electricity or used to irrigate huge areas of farmland. Salination is bound to follow.

How to see the good in the trees

There's an even easier way to create a desert - simply chop down all the trees. In northern temperate climates where the soil is deep this doesn't matter too much, but in other places trees are needed to keep shallow soil in place. Their roots hold the soil together and their leaves protect it from the weather. Just seventy years ago northern Ethiopia in Africa was a green and pleasant land. There were rich grasslands and forests with babbling streams where leopards roamed. Then the forests were all chopped down for fuel and to create more fields. The rain stopped coming, the soil blew away: what remains is a dry wasteland.

DEEP SOBS

Even small bushes and hedges are better than no trees at all. And even in the temperate north, soil will blow away from ploughed fields if it's not protected. There are roughly 600,000 kilometres of hedges left in Britain. They act as a windbreak and catch soil which may be blown from dry fields. But hedges are being uprooted fast. In places the soil is starting to go. Some farmers in the east of the country have started to plant plastic 'hedges' to try to hold on to their soil - how mad can you get?

JUST DUST

The ancient civilisations of Mesopotamia and the Indus took thousands of years to reduce their lands to desert. The Americans are moving much more quickly.

Big scale American-style farming makes for cheap food, but it's not good for the land. In the 1930s over-farming produced dust storms or 'black blizzards' of eroded soil in the American mid-west and thousands of farmers were driven from their farms. Today America loses *eight times* more soil every year than it creates. Eight billion tonnes of the stuff are blown away, into the rivers and out to sea. Already America may have lost a third of its top soil compared to two hundred years ago. The way things are going there won't be much soil left in America in few hundred years' time.

DUCKS IN THE MUCK

There are other ways to farm the land and to feed people. *Sustainable agriculture* aims to put as much back into the soil than it takes out in food. In some areas the best technique is the 'slash and burn' method of primitive tribes. Land which has been cleared of trees is left to regrow after the crops are harvested, while the farmers move on to clear a new patch of land. In modern countries this technique is impossible due to lack of space and time, but even so, if farmers look after their land the soil will last for ever - and even get richer.

Here's what they have to do:

Put compost and animal dung back on the land.

Plant hedges.

Give a field a rest every few years.

Grow several different crops and grow them on different fields in different years (this reduces the risk of pests).

This will protect farmland for the future.

See how Glacier Flint sums up.

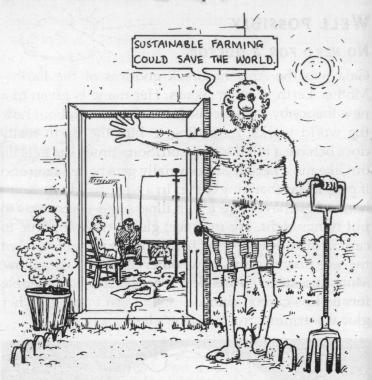

PARADISE!

WELL POSSIBLY

NO NEED FOR A DOCTOR

Gaia was the Ancient Greek goddess of the Earth - Mother Earth in other words. Her name is given to a new theory by the scientist James Lovelock about how the world works. According to him the Earth really does behave a bit like a person. Sometimes she gets ill, but like most people she normally gets better again. So if people are changing things in a way that doesn't suit her, Gaia behaves like it's an illness and takes steps to put things right. For instance: global warming due to carbon dioxide is heating her up, but if she gets too hot the seas may rise and many other disasters will follow. More disasters means less people and less people means less carbon dioxide 👣 - so bang goes her global warming problem. She's better again.

Or, of course, we may be due for another ice age, which would cancel out global warming!

Are you in harmony with Mother Earth?
- u hope you are!

If the Gaia theory is correct, then it's sensible to work with Gaia rather than against her: disasters such as floods are hardly the best way to keep life going. There have been many plans for people to live in harmony with the Earth and with each other. Such ideal societies are called *utopias* 👣.

The *Shakers* were an eighteenth century Christian sect, called Shakers because of their habit of whirling and shaking during church meetings. The Shakers did not care for large factories and towns. They believed that each of their small villages was a part of the Garden of

 Eden, the wonderful paradise where Adam and Eve lived before they sinned and God cast them out. Unfortunately Shakers aren't meant to have children so they have almost died out.

William Morris was a Victorian who hated modern industry. His utopia consisted of villages where free craftsmen farmed the land for themselves.

👣 *Utopia* is word invented by the English scholar Sir Thomas More (1478-1535). It means 'nice place'. More's Utopia was an island where, among other things, anyone who became too passionate about religion was sent away.

Findhorn is a spiritual community in northern Scotland. They grow incredible vegetables on soil which is little more than sand and gravel. They have been known to grow cabbages of up to eighteen kilograms. Findhorn is often quoted as an example of what can be achieved when people live in harmony with nature.

HANG IN THERE, MUM!

However, utopian communities on their own are unlikely to save humanity from further polluting the earth. Nowadays scientists suggest that we should *all* use less energy and have fewer children, and farmers should adopt sustainable techniques (see page 115). We just have to hope that Mother Earth will be patient with us until we get round to it - at least she has plenty of time.

She's probably got about six billion years left, not that anyone will be around to see her then. Mother Earth never stops changing. Just a few hundred million years from now the world we know will be unrecognisable. The continents will have moved, the oceans will have moved, climates will have changed, new species will probably have come along to replace us.

Towards the end of its existence, scientists think that the Sun will expand into a *red giant*, maybe expanding as far as Venus and even as far as Earth. But long before that time, life on Earth will have become impossible. The Sun's unimaginable heat will have turned the surface of Earth into a cauldron of boiling rock. Then finally the Sun will explode and all that will be left of the Sun, the Earth and the other planets will be a cloud of gas and dust - and who knows, from that cloud a new Earth and a new Sun may be formed.

It's all a long time ahead. Perhaps the best idea is just to enjoy living on this beautiful planet for as long as we can.

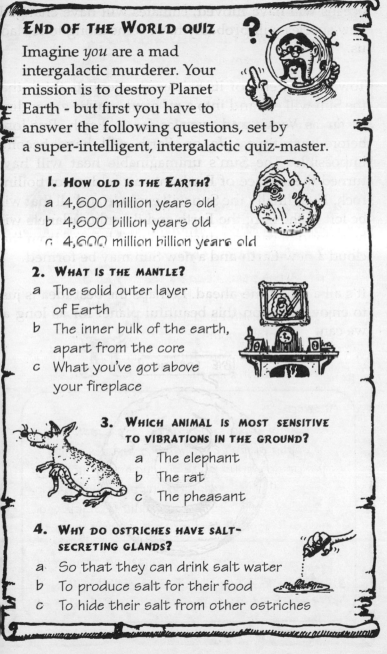

END OF THE WORLD QUIZ

Imagine *you* are a mad intergalactic murderer. Your mission is to destroy Planet Earth - but first you have to answer the following questions, set by a super-intelligent, intergalactic quiz-master.

1. HOW OLD IS THE EARTH?

a 4,600 million years old

b 4,600 billion years old

c 4,600 million billion years old

2. WHAT IS THE MANTLE?

a The solid outer layer of Earth

b The inner bulk of the earth, apart from the core

c What you've got above your fireplace

3. WHICH ANIMAL IS MOST SENSITIVE TO VIBRATIONS IN THE GROUND?

a The elephant

b The rat

c The pheasant

4. WHY DO OSTRICHES HAVE SALT-SECRETING GLANDS?

a So that they can drink salt water

b To produce salt for their food

c To hide their salt from other ostriches

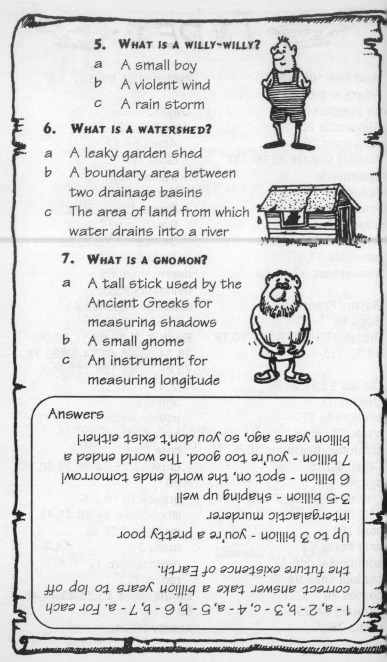

5. **WHAT IS A WILLY-WILLY?**

a A small boy

b A violent wind

c A rain storm

6. **WHAT IS A WATERSHED?**

a A leaky garden shed

b A boundary area between two drainage basins

c The area of land from which water drains into a river

7. **WHAT IS A GNOMON?**

a A tall stick used by the Ancient Greeks for measuring shadows

b A small gnome

c An instrument for measuring longitude

Answers

1 - a, 2 - b, 3 - c, 4 - a, 5 - a, 6 - b, 7 - a. For each correct answer take a billion years to lop off the future existence of Earth.

Up to 3 billion - you're a pretty poor intergalactic murderer

3-5 billion - shaping up well

6 billion - spot on, the world ends tomorrow!

7 billion - you're too good. The world ended a billion years ago, so you don't exist either!

INDEX

NOW READ ON

If you want to know more about Planet Earth, see if your local library or bookshop has any of these books.

THE STORY OF THE EARTH
By Peter Cattermole and Patrick Moore (Cambridge University Press 1985)
Everything you need to know about Mother Earth with loads of fascinating details - as well as an army of excellent diagrams and photographs. Finish this book and you'll be an Earth expert!

AROUND THE WORLD
By Steve Skidmore (Hodder and Stoughton 1992)
Join Jules Verne on an amazing journey round the world, plunge into the deepest mines and sail over the highest mountains.

DISASTER!
By Jim Hatfield (Watts Books 1994)
Find out the fearsome facts on Earth's worst disasters: the most violent volcanoes, the fiercest fires, the deepest floods - and much more! Don't read this book if you've got a weak stomach!

DEFEATING THE DESERTS
By Lawrence Williams (Evans Brothers 1989)
This book may cheer you up a little if the disasters have got you down. Find out why the deserts are spreading, but also how they could be stopped - there's hope for dear old Planet Earth!

ABOUT THE AUTHOR

Bob Fowke is a well-known author of children's information books. Writing under various pen names and with various friends and colleagues, he has created around fifty unusual and entertaining works on all manner of subjects.

There's always more to Fowke books than meets the eye - so don't be misled by the humorous style (just check out the index at the end of this book!). They're just the thing if you want your brain to bulge and your information banks to burble.

Bob Fowke is the youngest son of a Sussex vicar, and spent his childhood in the large, draughty vicarage of the village of Fletching (where the famous historian Edward Gibbon is buried). After years of travel and adventure, he now lives quietly in Shropshire.

OTHER BOOKS IN THIS SERIES

QUEEN VICTORIA, HER FRIENDS AND RELATIONS by Fred Finney. Watch out for the dumpy little lady in black.

HENRY VIII, HIS FRIENDS AND RELATIONS by Fred Finney. Too much meat, not enough toilets!

PIRATES OF THE PAST by Jim Hatfield. What life was really like beneath the Jolly Roger!

VILLAINS THROUGH THE AGES by Jim Hatfield. People not to take home to meet your parents!

SHAKESPEARE by Anita Ganeri. What the brilliant bard and his mad mates were really like.

ANCIENT EGYPTIANS by David Jay. They used monkeys to arrest burglars!

ELIZABETH I, HER FRIENDS AND RELATIONS by Bob Fowke. When men wore their knickers outside their tights ...

VIKINGS by Bob Fowke. Fancy a bowl of blood soup before bedtime?

THE ANGLO-SAXONS by Bob Fowke. Anyone for human sacrifice?

WORLD WAR II by Bob Fowke. Who won the War, and why men didn't have jacket pockets.

LIVING THINGS by Bob Fowke. From creepy crawlies to handsome humans - prepare to jump out of your skin!

SCIENCE by Bob Fowke. The secrets of the universe - and things which are even more interesting!

MUSIC by Nicola Barber. Funky monks make medieval music - and much more!

ROMANS IN BRITAIN by Bob Fowke. So why did they teach elephants to sit down to tea?

THE ANCIENT GREEKS by Bob Fowke. Handsome heroes and sun-baked beauties - but beware bird-bodied monsters, among other things!

D0120113

CANNABIS

Sarah Lennard-Brown

HODDER
Wayland

an imprint of Hodder Children's Books

White-Thomson Publishing Ltd,
2-3 St Andrew's Place, Lewes,
East Sussex BN7 1UP

Published in Great Britain in 2004 by Hodder
Wayland, an imprint of Hodder Children's
Books

This book was produced for White Thomson
Publishing Ltd by Ruth Nason.

Design: Carole Binding
Picture research: Glass Onion Pictures

British Library Cataloguing in Publication Data
Lennard-Brown, Sarah
   Cannabis. - (Health Issues)
   1. Cannabis - Juvenile literature
   2. Marijuana abuse -
   Juvenile literature
   I. Title
   362.2'95

ISBN 0 7502 4495 X
Printed in China by C&C Offset Printing Co., Ltd.

Hodder Children's Books
A division of Hodder Headline Limited
338 Euston Road, London NW1 3BH

Acknowledgements
The author and publishers thank the following for their permission to reproduce photographs and
illustrations: Corbis: pages 13 (Roger Wood), 21 (Ted Streshinsky), 23 (David Cumming; Eye
Ubiquitous), 27 (Jennie Woodcock; Reflections Photolibrary), 37 (James L. Amos), 51 (Stone
Les/Corbis Sygma), 53 (Roger Garwood and Trish Ainslie), 56 (Paul Hardy); Angela Hampton Family
Life Picture Library: pages 26, 43; Popperfoto: page 17; Rex Features: pages 9, 24 (Sipa Press), 29
(Bernadette Lou), 47 (David White), 48 (Ray Tang); Science Photo Library: pages 4, 7 (James King-
Holmes), 10 (Dave Reede/Agstock), 32 (Faye Norman), 34 (Jim Varney), 41 (Simon Fraser/Royal
Victoria Infirmary, Newcastle); Topham/ImageWorks: pages 30, 38, 44 (©Esbin-Anderson), 54
(©Skjold); Topham/PA: cover and pages 1, 45, 58; Topham Picturepoint: pages 16, 19. The
illustration on page 28 is by Carole Binding.

Note: Photographs illustrating the case studies in this book were posed by models.

Every effort has been made to trace copyright holders. However, the publishers apologise for any
unintentional omissions and would be pleased in such cases to add an acknowledgement in any
future editions.

Contents

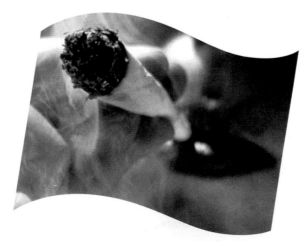

Introduction
A drug with a difference

The United Nations Office on Drugs and Crime estimates that, worldwide, 180 million people over the age of 15 used drugs recreationally at some point during the late 1990s. This is about 4.2 per cent of the world's population. Out of the total who had taken drugs, by far the largest group – 144 million people – had taken cannabis. The next most commonly taken drugs were the amphetamine-type stimulants, with 29 million users worldwide. By contrast, cocaine had only 14 million users worldwide.

What is cannabis?

Cannabis is a plant that has been used in many ways for thousands of years. Its fibres have been used to make rope, paper and cloth. Cannabis seeds have been food for people and animals. The plant has also been used to make medicines. In addition, people discovered that smoking or eating some parts of the plant has an intoxicating effect. Cannabis users say that it makes them feel relaxed, happy and stimulated. This effect comes from chemicals that are present in the plant. They affect the user's brain and nervous system. There is more about the history of cannabis in Chapter 1, and about its effect on the mind and body in Chapter 2.

An amazing plant, a complex argument

The cannabis plant is useful in many ways, but its psychoactive qualities have given it a chequered past and an uncertain future.

4

Two types of cannabis plant

Over the centuries that people have used the cannabis plant, it has been selectively bred to improve its usefulness. Today there are two main varieties.

Cannabis sativa plants are tall, growing up to 6 metres high, and can be planted very close together. They are also resistant to many common plant diseases. This means that cannabis sativa gives a high yield (quantity of crop) per hectare of land, and the fibres produced by the plants are very long, which makes them suitable for spinning and weaving. The concentration of the chemicals that produce the intoxicating effect of cannabis is so low in cannabis sativa that this variety of the plant is not considered to be 'psychoactive' (affecting the brain and producing intoxication). Cannabis sativa tends to be grown by people wishing to use the plant in industrial processes.

Cannabis indica is a small shrubby plant, which produces a high concentration of the chemicals that produce intoxication. People wishing to use the plant as an intoxicant usually grow cannabis indica.

A controversial drug

For most drugs that people take recreationally, there is no dispute about the damage they do to the user and to society. For example, no one argues that crack cocaine is not addictive and that people should be free to use it. Crack cocaine is so addictive that users often resort to crime to obtain money to buy the drugs they crave. The drug itself can make users violent. However, in the case of cannabis, there is much debate. Some people argue that cannabis is less harmful than tobacco and alcohol, and does not cause users to become violent or so addicted that they commit crimes in order to obtain money to buy the drug. They feel that the law should be changed so that adults could use cannabis if they wished. A different point of view is that cannabis is a 'gateway drug', introducing people to the world of illegal drugs and leading them on to abuse drugs that are more harmful. People who see cannabis in this way argue that it should remain an illegal drug.

Where cannabis plants for drugs are grown

From the mid-twentieth century, many countries passed laws designed to stop people using harmful recreational drugs. These included laws prohibiting the growing of cannabis. As a result, the number of countries producing illegal drug crops has come down. However, 130 countries still reported illegal cultivation of cannabis to the United Nations Office on Drugs and Crime during the 1990s.

Cannabis grows better in some climates than others. *Cannabis indica* grown in bright sunlight produces the highest concentration of intoxicating chemicals. Some of the strongest preparations of the drug are produced by growing the plant indoors, under controlled conditions. However, this method of growing is very expensive and so most illegal cannabis is grown where the climate is most suitable. The greatest amount of illegal cannabis is grown in South America and Asia.

United Nations

The United Nations Office on Drugs and Crime was set up in 1997 to coordinate the international fight against drugs and crime. It collects information about drug production, drug smuggling and drug use around the world. It advises governments about making laws to combat drug abuse in their countries. It is also involved in investigating and prosecuting drug-related crime and terrorism.

The cannabis trade

From the places where it is produced, cannabis is smuggled into countries around the world. This trade is illegal and customs officers and police try to stop it. Most of the cannabis seizures they make occur in just a few countries. Spain, the UK, Pakistan, the Netherlands and Morocco are the countries where three-quarters of all cannabis resin seizures take place. Mexico, the USA, South Africa, Colombia and India are the countries where three-quarters of all cannabis herb seizures occur. Cannabis herb accounts for the largest amount of illegal drug traffic around the world, followed by cannabis resin, then cocaine.

Cannabis laws today

Ever since cannabis was made illegal, some people have

argued that the laws should be changed. In trying to stop cannabis being used as a recreational drug, the law-makers had also prevented the cannabis plant from being used in the manufacture of several useful products. Some people say that cannabis is a helpful medicine for treating or relieving the symptoms of some serious illnesses. These arguments, as well as the argument that cannabis as a recreational drug is not harmful, have been put forward by people who campaign for cannabis use to be legalized.

Over time, some countries have relaxed their laws on cannabis, and the law about cannabis has become a 'hot' political issue. Chapter 3 looks at the potential uses of cannabis in today's world, and Chapter 4 investigates how different countries try to control the use of cannabis among their population. It also presents the arguments that people put forward, on both sides.

Cannabis and you

With the law on cannabis being a popular topic for discussion and campaign, it is likely that you already have some ideas about the drug. You may know people who have used the drug, or are regular users. You may have been invited to try cannabis yourself, or this may happen fairly soon. The purpose of this book is to give you the information you need to make up your own mind about the issue and to make the healthiest choice for yourself.

Being there
It can be great fun being part of the scene and hanging out with friends, but it must always be your own decision when it comes to taking substances that could affect your health.

1 The history of cannabis
A plant with many uses

The cannabis plant probably originated in central Asia, where the ancient nomads found its tough, fibrous stems so useful that they took its seeds with them when they moved around. Traces of cannabis plants have been found as far apart as in the tombs of the ancient Egyptians and in China, where cannabis has been grown as a crop since people started to settle in that part of the world. One of the earliest records of people using the cannabis plant has been found in Taiwan, an island close to China. Archaeologists found a clay pot that had been decorated with cord made from cannabis fibres. This Stone Age pot was made over 10,000 years ago.

The ancient Chinese

The ancient Chinese found that fibres twisted together were stronger than single fibres and so they started to produce rope. The strong, long fibres they needed came from the cannabis plant. They also began to weave the fibres together to produce cloth. China is famous for its silk. Silk is woven from the fine threads produced by silkworms, but these are very expensive and so silk could only be afforded by the rich. Cannabis fibres (called hemp) were far cheaper to produce and cannabis became such an

Useful history
The long stems of cannabis have been used since earliest times for their fibre.

important crop that the ancient Chinese referred to their country as 'the land of Mulberry and Hemp' (silkworms feed on mulberry leaves).

The Chinese realized that the cannabis plant had multiple uses. They discovered that the male plants produced better fibre for rope and cloth, and the female plants produced seeds, which were a nutritious food for animals and people. The Chinese found that the strength and flexibility of cannabis fibres made them suitable for producing high-quality bowstrings for their archers. These bowstrings were so important in maintaining China's military superiority that large areas were set aside to grow cannabis exclusively for this use.

One of China's many great innovations was the invention of paper. Before paper, the Chinese wrote on slips of wood, which were very heavy. Silk was used occasionally for special documents, but this was very expensive. The first evidence of paper dates back to the first century BC. Cannabis fibre (hemp) was the main fibre in paper and continued to be used across the world for nearly 2,000 years. The first drafts of the American Declaration of Independence, in 1776, were written on hemp paper.

Paper

The Chinese first made paper by soaking ground-up mulberry tree bark and cannabis fibres in a tank of water. The fibres were then skimmed from the top of the tank in moulds and left to dry. The Chinese closely guarded the secret of making paper. The first people to discover this secret were Arabs and they started Europe's first paper mill in 1150.

The ancient Chinese also investigated the medical properties of cannabis. From at least as long ago as the 28th century BC, they used cannabis preparations to treat menstrual problems, rheumatism, constipation, gout and many other conditions. It was also used as a pain reliever. Later the intoxicating effects of cannabis were used in religious settings to help people 'communicate with the spirits'.

Japan and India

Cannabis production is woven through the history of many great civilizations, including Japan and India. In Japan, like China, cannabis was valued for its fibre and commonly used in the production of cloth. It also became an essential component of medicines and religious ceremonies.

In India, one of the Hindu gods, Shiva, is said to have brought the cannabis plant down from the Himalayan mountains for 'use and enjoyment'. The earliest written reference to the psychoactive (mind-changing) properties of cannabis is found in an ancient Hindu text, called the *Atharva-veda*, which is thought to date from approximately 2000 BC. Unlike in the Far East, cannabis became popular as an intoxicant and was often prepared as a concoction, mixed with other herbs and spices, and used in a similar way to alcohol in the West. Different varieties of concoction were called bhang, charas and ganja and were used to celebrate weddings and births, as symbols of hospitality and welcome, to fortify soldiers in battle and to aid communication with the gods in religious ceremonies. Cannabis was also used as a food (hemp seeds) and as a medicine to treat dysentery, sunstroke and digestive disorders.

The Buddha

According to the Mahayana Buddhist tradition, the Buddha is supposed to have survived on one hemp seed a day during a long fast.

Ancient Greeks and Romans

The ancient Greeks recognized the many uses of the cannabis plant. It is referred to in *The Histories* written by Herodotus (c. 485-425 BC). The Greeks did not use the plant for its intoxicating properties, but valued it as a source of fibre. In the 6th century BC, the Greeks were carrying on a thriving trade in hemp fibre. They also used the seeds as a treatment for backache and as a food.

Hemp rope was very important for the Roman Empire, especially to equip Roman ships, and large quantities of cannabis were imported, mainly from Babylon. The Romans also recognized the medicinal uses of cannabis, and cannabis is mentioned in the first Western directory of

healing plants, *Materia medica*, written by Dioscorides in AD 70. The famous Roman doctor Galen (AD 130–200) wrote about cannabis as a cure for earache and gout, but he noted that it could cause impotence in men if over-indulged in.

Roman rope

The rope used on ancient Roman ships was made from cannabis fibres.

Europe and America

Europe was persuaded of the value of hemp rope and fibre. In 1563 Queen Elizabeth I of England passed laws that required owners of farms above a certain size to grow cannabis. The crop was so important, especially for rope for ships, that some English towns (Hemel Hempstead, for example) were named after it.

'We still use hemp rope today. We use it on our sailing boat. It's very strong.' (Ben, aged 14, sailor)

Cannabis was introduced to the American continent by Spanish sailors in 1545, and it became an essential crop for the New World. In 1619, a law was passed at Jamestown Colony in Virginia, requiring all farmers to grow cannabis for fibre.

Cannabis Timeline

BC

10000	Cannabis cord used to decorate a Stone Age pot in Taiwan.
8000–7000	Earliest evidence of cannabis fibre cloth, in China.
2700	Cannabis listed in the pharmacopoeia (list of medicinal plants) of Shen Nung (China).
2000	Psychoactive properties of cannabis mentioned in the ancient Hindu text, the **Atharva-veda**.
550	Cannabis listed in the **Zend-Avista**, a pharmacopoeia written by Zoroaster, a Persian prophet whose followers are credited with introducing the plant to India.
5th century	Greek historian Herodotus records the use of cannabis in **The Histories**.

AD

1st century	Chinese start making paper from hemp fibre.
70	**Materia medica**, a pharmacopoeia by Dioscorides, includes cannabis.
6th century	Greeks organize a profitable international trade in hemp fibre.
1150	First European paper mill using hemp fibre is built by Muslims at Xativa, in Spain.
1545	Cannabis is introduced to America.
1563	Queen Elizabeth I of England orders landowners to grow cannabis.
1564	King Philip of Spain orders cannabis to be grown throughout his empire (from Argentina to Oregon).
1619	Legislation in Jamestown Colony, Virginia, orders farmers to grow cannabis.
1776	First two drafts of the American Declaration of Independence are written on hemp paper. Betty Ross sews the first American flag, using hemp cloth.
1798	French emperor Napoleon and his army invade Egypt, where they learn about smoking cannabis as a drug. When they return to France in 1801, they take the practice back with them.
1840	Club des Hachischins is established by Bohemians in Paris.
1890	Queen Victoria's doctor, Sir Russell Reynolds, writes of prescribing cannabis for menstrual cramps.

1924	Second International Opiates Conference includes cannabis in a list of dangerous drugs whose production and use should be reduced.
1925	Dangerous Drugs Act in the UK makes it illegal to import, export, process, produce, sell or buy any drug on a list that includes heroin, cocaine and cannabis.
1931	Marijuana Tax Act in the USA effectively stops the production of hemp fibre.
1941	Henry Ford creates a 'hemp' car, made from and powered by cannabis.
1942	Restrictions on growing cannabis are lifted across the USA, in order to supply the need for fibre and cordage during the Second World War.
1951	Narcotics Control Act in the USA introduces heavy penalties for possession of any cannabis product.
1961	United Nations Treaty 406 Single Convention on Narcotic Drugs: 40 countries agree to work towards eradicating the production and consumption of opium, cocaine and cannabis.
1971	Cannabis classified as a Class B schedule 1 drug under the Misuse of Drugs Act 1971 in the UK.
1974	Netherlands relaxes its drug laws to allow the sale of small amounts of cannabis, which must be used at regulated premises.
2000	The US state of Hawaii passes legislation to enable cannabis to be prescribed as a medicine for certain life-threatening illnesses. Hawaii is one of ten states that allow the use of cannabis as a medicine in some situations.
2003	Cannabis is downgraded from a Class B drug to a Class C drug under the Misuse of Drugs Act in the UK. This means that, although it remains an offence to be found in possession of cannabis, the penalties are reduced from 14 years' imprisonment to 2 years – and, in most cases, people found with very small quantities of cannabis for personal use will be given a caution, rather than prosecuted.

Discovering the drug

By the nineteenth century cannabis fibre was used all over the world to make rope and sacking, but the plant's intoxicating effects were not known in Europe. This situation changed after the French Emperor Napoleon invaded Egypt in 1798. Napoleon's soldiers learned about cannabis as a drug from the Egyptians and took the discovery home to France. Nineteenth-century European scientists then investigated the new drug and this led to its becoming a popular remedy for many ailments. Sir Russell Reynolds, the medical advisor to Queen Victoria (1819-1901), prescribed the plant for insomnia (inability to sleep), nausea (feeling sick) and menstrual cramps. It was possible to buy cannabis preparations in chemists' shops without a prescription until the 1890s.

Club des Hachischins

The Club des Hachischins was started in France in 1840 by a group of intellectuals, artists and writers who met to talk and experiment with cannabis. The idea was that cannabis helped to stimulate their artistic and intellectual thinking. Members included the poets Charles Baudelaire and Arthur Rimbaud and the poet, novelist and painter Pierre Gautier. This group were part of a larger group known as Bohemians, who congregated in the cafés around Montmartre in Paris. They rebelled against the social conventions of their time, and held radical views on morality, art and politics. Famous Bohemians include the writer Victor Hugo and artists Jean-François Millet and Gustave Courbet.

Paris scene
Cafés in Paris became the centre of the 'Bohemian' movement in the second half of the nineteenth century.

Cannabis cultivation in the twentieth century

The amount of cannabis grown across Europe and America decreased at the end of the nineteenth century and the beginning of the twentieth century. Other crops were used for animal feeds, and new synthetic (manmade) fibres were beginning to become available, replacing hemp. However, during the First and Second World Wars, the need to grow cannabis increased, in order to produce hemp fibre for making rope, sacking and cloth. During the Second World War, in Britain, the USA, Canada, Australia, France and particularly in Germany, growing cannabis was seen as a way of supporting the national war effort. Before, these countries had been getting the hemp they needed from Russia, but the supply was now cut off because Russia was at war with Germany. To meet their increased need for hemp fibre, each country vastly increased its cultivation of cannabis. In 1943 the USA issued a 'pro hemp' propaganda film called *Hemp for Victory*, in which farmers were shown how best to cultivate the crop and the many uses to which it could be put. However, as soon as the war was over, this policy was revoked.

War effort
Growing cannabis was encouraged during the Second World War. One important use of the fibre was to make sacking for sandbags.

Cannabis becomes an illegal drug

As cannabis ceased to be so important as a crop, there began to be increasing publicity about the intoxicating effects of cannabis, and there were calls for laws to prevent people using it. During the late nineteenth and early twentieth centuries, concern grew about the effects of intoxicating substances on the general population. In Britain there were worries about the number of people abusing laudanum (a form of opium which could be purchased without prescription from chemists) and also alcohol. In the USA there was increasing concern about the effects of all intoxicating substances, particularly on working people. These substances were seen as the cause of many social problems and it was felt that people needed to be protected from their evil influence.

Several international conferences were held to investigate the effect of intoxicating substances on the general population. In 1924, the Second International Opiates Conference at the Hague in the Netherlands looked not only at opiates (drugs such as heroin and morphine, produced from the sap of the opium poppy) but at all intoxicating drugs. The Conference produced an agreement to control and reduce the production and use of narcotic drugs across the world, and cannabis was included in the list.

This was the beginning of a worldwide campaign against recreational drugs. In the UK, cannabis was declared an illegal drug, along with heroin and cocaine, in the Dangerous Drugs Act 1925. The law made it an offence to grow, process, distribute, sell or possess cannabis or related products. This caused the production of all varieties of hemp to be stopped in the UK. An era of prohibition followed around the world, during which it was illegal to produce, sell or buy intoxicating drugs (including alcohol in the USA). However, making a

Violent connection?

During the 1920s and 1930s newspapers portrayed cannabis as causing violent behaviour. Research into cannabis and violence has not found any connection between the two.

Prohibition

During Prohibition in the USA (1920–33), it was illegal to produce, sell or buy intoxicating drugs, including alcohol. People found ways of hiding alcohol to use in secret. Some people argue that laws to discourage the use of drugs actually make them more attractive and increase their use.

substance illegal does not entirely stop people wanting to take it or being addicted to it, and so drug production and supply became a profitable secret occupation for criminals and smugglers.

The anti-drug campaign was taken up enthusiastically by newspaper editors and several influential people in the USA, including a man called Henry Anslinger who became head of the Federal Bureau of Narcotics in 1930. He started a campaign against all drugs, in particular cannabis, which he felt made people (especially those of Afro-Caribbean origin) violent and lazy. There were protests from the hemp industry, who saw their livelihood being made illegal, and from some doctors, who felt that the reasons given for criminalizing cannabis were racist and unfounded. Nevertheless, in 1937 the Marijuana Tax Law was passed, which stated that anyone selling cannabis (whether for use as a drug or as fibre) had to pay a 'transfer tax'. Under this law, you could only apply for a 'tax stamp' when you were in possession of cannabis – but being in possession of cannabis without having paid the stamp was an offence. This meant that it was impossible to be legally in possession of cannabis, and so the Tax Law effectively made the production of hemp fibre illegal. Further laws in the USA, including the Narcotics Control Act of 1951, introduced heavy fines for people found in possession of, selling or smuggling cannabis, whether it was for use as a drug or as a fibre. This eliminated the production of what had been a staple crop across the USA. By 1950, farming of cannabis had been banned across the whole of the USA.

Henry Ford and the chemurgical movement

One of the people who thought that cannabis was useful was the industrialist Henry Ford. He was a supporter of the Farm Chemurgic Council, which was formed in 1935 by people who believed that the normal crops grown by farmers could be used to develop unlimited new products and technologies. At this time, the developments of the petrochemical industry (producing plastics, nylons, dyes and paints, etc) were in their infancy. Supporters of the chemurgical movement argued that all these products could be produced just as well from plants as from petroleum, and that plants had the added advantage that they would not dry up like oil wells.

In 1941 Henry Ford was investigating how crops could be converted into fuels and materials for the automobile industry. He was very keen on the use of ethyl alcohol, made from fermented hemp and other vegetation, as a fuel. He felt this would be ideal to power machinery, including his new cars. He also produced a prototype car with a strong and robust body made from hemp fibre strengthened with resin. But, unfortunately for Henry Ford, the tide of opinion against all forms of cannabis meant that his plans for the car had to be abandoned. Today, companies such as Mercedes Benz and Daimler use a lot of hemp fibre in their factories and there are plans to use hemp fibre and resin in car bodies, just as Henry Ford suggested in 1941.

A symbol of rebellion

With hindsight, it can be seen that making cannabis illegal did little to reduce its use as an intoxicant. Some people argue that the laws actually encouraged more people to try the drug, as it became a symbol around which rebellious groups could rally. The first group to use cannabis in this way were the 'beat generation'. This group, who are identified with Jack Kerouac, Allen Ginsberg and Ken Kesey's 'Magic Bus', rebelled against the social conventions of 1950s' America and sought freedom and self-expression. In reality, this meant that they were promiscuous and took a lot of drugs. However, their idea of exploring your own mind and realizing your artistic potential with the aid of illegal narcotics (including cannabis) created a sense of romance and excitement. This was very exciting for a generation of young people who had been brought up during the hardships of the Second World War.

Merry Pranksters

The writer Ken Kesey (1935-2001) was a leading figure in American counter-culture, known for using cannabis and other drugs. His followers became known as the Merry Pranksters and in the 1960s they made a famous journey to New York in his 'Magic Bus', painted in psychedelic colours.

The 1960s and into the present

Cannabis use as a drug really became popular during the 1960s. It began to be part of the popular youth scene and was taken up by the Hippy movement. The 'Summer of Love' in 1967 saw a surge in the use of cannabis as an intoxicant in the Western world. Several film and pop stars were arrested for possession of cannabis.

Cannabis was also closely associated with the Caribbean island of Jamaica and with the Rastafarian religion, which became popular there during the 1930s. Cannabis was probably introduced to the island by Indian labourers in the nineteenth century. The common Jamaican name for it is 'ganja', which comes from India. Cannabis is used by Rastafarians as part of their religious ritual and as a medicine. Rastafarians feel that the laws that prevent

'In 1967 we were so full of hope for a new world. There was lots of talk about drugs, but I never took any. I didn't really see any. I wouldn't have known what to do with them. I think there is more cannabis around today than there was then.' (Marion, aged 57, ex-Hippy)

The Rastafarian religion

Rastafarianism is a religion and a way of life. Its religious book is the Bible, but Rastafarians interpret it differently from mainstream Christians. They see white political power as 'Babylon', an evil oppressive power that keeps Rastas in bondage. One of the prime beliefs of the Rastafarian religion is that Emperor Haile Selassie I of Ethiopia was the living God of the black race. (The name Rastafarian comes from Haile Selassie's original name, Tafari, preceded by 'Ras', meaning 'prince'.) When Haile Selassie was murdered in 1975, many Rastafarians refused to believe that he was dead. They now believe that he 'sits on the highest point of Mount Zion', awaiting the time of judgement.

Rastafarians consider Africa to be 'heaven on earth'. They believe that Jah (God) will send a signal and that they will travel back to Ethiopia. This is called the 'exodus', like the Bible story of the Israelites escaping from slavery in Egypt and entering the 'promised land'.

Rastafarians
One distinctive feature of Rastafarians is that they wear their hair in long dreadlocks. Cannabis is used as part of some Rastafarian religious ceremonies.

them using cannabis in their religious rituals are wrong (cannabis is illegal in Jamaica) and they continue to campaign to be allowed to continue using it.

As the twentieth century progressed, cannabis was the subject of much debate. Individuals and groups campaigned to be allowed to use it as a medicine and as an industrial material, and some people argued that the law should be changed so that cannabis could be used as a recreational drug. In Chapter 3 we will look at how cannabis is used in today's world, but first we will explore the effects that the drug has on people who use it recreationally.

2 Taking cannabis The effects on mind and body

People who take cannabis recreationally say that it makes them relaxed and happy. They feel a mix of tranquillity, hilarity and drowsiness. In general, people who smoke cannabis feel stimulated and happy at first. These feelings are followed about half an hour later by drowsiness.

Cannabis is unique in the drugs world in that it seems that the brain may have to learn how to respond to it. Sometimes, the first time a person takes cannabis, they hardly notice any effects. If they take it again, the effects become more noticeable.

How the intoxicants enter the body

When a person smokes cannabis, hundreds of chemicals in the smoke, from the cannabis and the tobacco, are quickly absorbed by their lungs. Within minutes, the chemicals enter their bloodstream and are taken to their heart, brain and other organs. The most psychoactive chemical in cannabis smoke (the one that affects the brain the most) is called delta-9-tetrahydrocannabinol – THC for short. The huge number of chemicals contained in cannabis have not been fully researched, and so it is still unclear whether they react with each other to produce the effects and how exactly they cause the effects. Most research has concentrated on the effects of THC.

Taking in THC
Cannabis smoke contains the psychoactive substance THC, or delta-9-tetrahydrocannabinol. It quickly passes via the user's lungs into their bloodstream.

Most of the THC that reaches the brain is broken down and removed by the body's natural mechanisms within a few hours. However, other organs, such as the kidneys, liver, spleen and testes, do not get rid of THC as fast as the brain. In the liver, some of the THC is converted into compounds that stay around for several days. Some of these compounds also have psychoactive effects, so, although the first effects of cannabis are over within a few hours, the drug continues to be present in the bloodstream and to affect the brain and body for several days.

THC is taken up by body fat, which acts as an energy store for the body when there is not enough food to meet its energy needs. When the body fat is broken down to release the energy, the THC is also released into the bloodstream again. So, even if someone has not taken cannabis for several weeks, it may still be detectable in their body and may still affect their brain and body.

Immediate effects

The effects of smoking cannabis vary, depending on the concentration of intoxicating chemicals it contains. When high-grade cannabis is taken, the effects are felt less quickly, but when they do occur, they can be almost hallucinogenic (causing the person to see, feel, smell or hear something that does not exist). The person may experience a heightened awareness of some colours or smells and of their own body. They may be very aware of their heart pumping and the blood flowing around their veins, or their muscles tensing and relaxing. This can be frightening and cause the person to panic. Sometimes they attend hospital because they are so worried. However the best treatment is quiet, calm reassurance. The sensation will diminish, in half an hour up to several hours.

Low-grade cannabis has very little effect. Sometimes it is referred to as 'headache pot', because the user tends to get a feeling of heaviness in their head or a headache, rather than a 'high'.

Eating cannabis

When cannabis is eaten, larger amounts tend to be taken, and so the effect lasts for longer and can be more intense.

The immediate effects of cannabis depend on the individual. Many people do not experience anything the first, or even the first few times they take the drug. How experienced users feel when they take cannabis also varies. Some feel that they understand the world better; others feel happy. The feeling people get when they take the drug often depends on the environment they are in and the people they are with. Sometimes the experience is not very good.

The effects of taking cannabis on the body are easier to identify. The user's eyes become bloodshot and their face may become flushed red. They may feel their heart beating more quickly. Taking cannabis also produces an effect called 'the munchies' – an intense feeling of hunger that is difficult to satisfy.

People often have difficulty recalling what it was like when they took the drug, and this may be due to its effect on memory. Taking cannabis disrupts the ability to remember things. Users often report that time seems to slow down when they take cannabis: a few

I'm worried about my brother

'My brother smokes pot occasionally. I don't think my parents notice, but I do. His eyes get all red and he eats everything – all the biscuits and crisps. He says he does it because his friends do it. Rubbish reason! He's a bit pathetic, my brother. I don't think the pot has much effect on him really – he always sleeps a lot anyway. I haven't told anyone. I'm worried what Mum and Dad will say if he gets caught. Someone told me that, if the police catch him with cannabis here, my parents will be in trouble and have to go to court. He just laughs and says no one is going to catch him. I don't know what to do. All he thinks about is himself.'
(Andrew, aged 15)

minutes can seem to take an age. This may also be due to the effect of cannabis on memory and concentration. These feelings can last at least 48 hours after taking the drug and can recur up to three weeks later, as the last traces of THC are removed from the body.

Immediate effects
Noticeable effects of taking cannabis are a flushed complexion and bloodshot eyes.

Short-term and long-term effects of taking cannabis

Short-term effects	Long-term effects
Relaxation	Memory formation and learning are
Sedation	impaired for 48 hours or more.
Hilarity	Functioning of immune system is
Tranquillity	probably affected.
Lack of aggression	Lung disease (probably related to
Impaired judgement	smoke inhalation).
Increased heart rate	High doses can result in decreased
Flushing	sperm count for males and
Blood-shot eyes	irregular periods for women.
Inhibition of memory	Brain development of unborn
Inhibition of ability to learn	child can be affected if cannabis
Hunger ('munchies')	is smoked during pregnancy.

The basal ganglia, underneath this structure, controls coordination of movement, expression and emotion.

The hippocampus is involved in making and storing memories.

The brain stem does not have any cannabis receptors. This area regulates breathing.

The cerebellum controls coordination of movement.

Cannabis and the brain

One fascinating thing about cannabis is that there appear to be specific areas in the human brain that recognize and respond to THC. They have been called the cannabinoid receptors, and they are found in several areas of the brain, including the hippocampus, the cerebellum and the basal ganglia. There are no receptors for THC in the brain stem, an area that controls breathing. The lack of THC receptors in this area may explain why an overdose of cannabis is not fatal.

Cannabis is not the only plant for which there are specific brain receptors. Our brains also have opiate receptors, which respond to opiate drugs such as heroin and morphine. It is known that our opiate receptors exist because we naturally produce a chemical similar to opiates, called an endorphin, to regulate pain and stress within our bodies. Therefore, it would be logical to think that, since our brains have THC receptors, we must also naturally produce a chemical that is similar to THC.

There are two main neurotransmitters (chemicals) in the brain that stimulate the cannabinoid (THC) receptors. They are anandamide and 2-AG. What these neurotransmitters do and how they work is not yet completely understood,

The brain

There are cannabis receptors in the areas of our brain known as the basal ganglia, the hippocampus and the cerebellum. Each area controls different functions.

but it would seem that anandamide has something to do with feelings of tranquillity and 2-AG decreases the ability of the brain to use memories efficiently. Why the brain should produce a chemical that reduces the ability to remember is not clear. It may be to restrict the ability to remember things that are very painful or damaging. These effects of anandamide and 2-AG can both be observed when people use cannabis. In fact, the effects of cannabis on memory and the ability to create new memories mean that taking the drug can result in poor performance at school, work or play.

'My mate Glen dropped out of college. His teachers said he wasn't trying hard enough, but really it was drugs. He got in with a group of guys who smoked joints. He got all laid back and just didn't care about anything.' (Mark, aged 20, student)

Some of the cannabinoid receptors are in the cerebellum and basal ganglia. These parts of the brain are involved in the coordination of movement and the judgement required in making fine movements. Cannabis is similar to alcohol in the way it disrupts the user's ability to control movement, making it dangerous to drive or operate machinery.

Keeping alert

To cycle safely in traffic, you need to be in control of your movement, well coordinated and alert to everything that is happening around you. Cannabis has a negative effect on all those factors.

The majority of our cannabis receptors seem to be located in the hippocampus. This part of the brain is involved in making and storing memories. If your hippocampus is damaged, you have problems learning new tasks, but you can remember things you learned in the past.

Scientists researching the effects of THC have shown that, when rats are given THC, activity in the hippocampus is reduced. It returns to normal as the THC wears off. Other rat and primate experiments have shown that, under normal conditions, THC does not seem to produce long-term damage to the structure of brain cells. Only if the doses are much higher than a human would normally be able to take does some damage to brain cells occur.

Some scientists are cautious about this. They feel that THC may change the way brain cells connect with each other, or may change the concentrations of the chemicals in the brain that enable brain cells to communicate. These sorts of changes are very hard to show when a brain is examined under a microscope. However, some studies of people who have used cannabis for many years have shown that there may be difficulties with memory and problem-solving. This needs further study to be sure that the difficulties are not due to THC present in the individuals at the time of the test, rather than being due to long-term heavy cannabis use.

Cannabis and the heart

Cannabis has the effect of increasing the user's heartbeat. They may feel their heart is racing, as it would after running or when they are very excited or nervous. The

Heart rate
Cannabis makes the user's heart beat much faster than normal. The feeling is the same as when you take very strenuous exercise.

increase in the heart rate can be as much as 20 to 30 beats per minute. An average adult has a heart rate of 60 to 70 beats per minute, so raising this to 90 or 100 beats per minute is quite a big rise. However, it is quite normal for the heart rate to rise to as much as 120 or 130 beats per minute during strenuous exercise, with no bad effects on the person's health. So, it would seem that the kind of rise in heart rate caused by cannabis would not be dangerous for someone in good health, with an average normal heart rate. It could be dangerous for someone whose heart rate is already higher than average, due to high blood pressure or heart disease.

'My uncle has smoked pot for years. He tells everyone that it doesn't do him any harm, but he's got a cough like he smokes 40 a day.'
(Cassie, aged 15)

Cannabis and the lungs

Smoking cannabis does affect the lungs. Often the damage caused by smoking cannabis is worse than that caused by smoking normal cigarettes. This is due to several factors. Firstly, people smoking cannabis construct their own cigarettes. Unlike many manufactured cigarettes, these do not have filters, which do help a little to reduce the amount of tar and chemicals that the smoker inhales. Secondly, to make sure that the maximum amount of the drug enters their system, cannabis users tend to take bigger breaths of smoke deep into their lungs and hold it there for longer than people who smoke cigarettes without cannabis. This magnifies the damaging effects.

Tobacco smoke and cannabis smoke contain similar chemicals that damage the cells lining the lungs. Therefore people who smoke plain cannabis are at a similar risk to those who smoke a joint of cannabis and tobacco. Lung, throat and mouth cancers are known to be a risk for people who smoke tobacco. It is not known whether smoking cannabis can cause these cancers, but anyone smoking cannabis mixed with tobacco is clearly at risk.

Asthma

Cannabis can trigger an asthma attack for some people with asthma. They need to avoid smoking cannabis and to stay away from places where cannabis is smoked.

Cannabis and the immune system

There are THC receptors all over the body, not just in the brain. Some are found in the cells that control your immune system, by which your body protects itself from infectious illness and disease. Only few studies have looked at this area. They show that cannabis may decrease the body's ability to fight infections. More work needs to be done before we can be sure.

Cannabis and the reproductive system

Using cannabis seems to affect the hormones that regulate sperm production in men and egg production in women. It does this by increasing the production of a hormone called prolactin. Two other effects of this on men may be difficulty in achieving and maintaining an erection and the growth of breast tissue. In women, the hormone disruption can result in irregular periods.

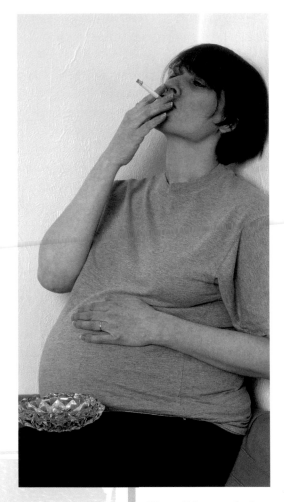

Taking cannabis during pregnancy

Several studies have shown that babies born to women who used cannabis during pregnancy tend to weigh slightly less than other babies. This may be due to the cannabis being smoked with tobacco, since tobacco is also known to cause low birth weight in babies. In a study by Italian scientists in 2003, pregnant rats were injected with cannabis extracts. The baby rats that were born were found to be more 'hyperactive' than normal and scored lower on learning tests throughout their lives. The scientists feel that this proves that cannabis can pass to the unborn child from the mother. Their work supports other studies showing that children born to mothers who took cannabis while pregnant may have long-term memory and learning problems.

Smoking risks

Smoking tobacco while pregnant carries risks to the unborn baby. Mixing cannabis with the tobacco adds further dangers.

Cannabis and mental health

The effect of cannabis on mental health is hotly debated. Some people think there are links between cannabis use and depression and between cannabis use and schizophrenia, a disease of the mind that involves thought disturbance, bizarre and unusual behaviour and loss of understanding of reality. Percentages of cannabis users with depression and schizophrenia are greater than the percentages of non-cannabis users with these disorders. However, this does not necessarily mean that cannabis causes the disorders. It could be that lifestyle, genetic and family influences make people with depression and schizophrenia more likely to have the opportunity to use cannabis and more likely to become dependent on it.

Although cannabis use has increased over the last half century, the proportion of cannabis users who have mental health problems has not increased. Some mental health researchers feel that people with certain mental health problems may use cannabis, like alcohol, in a mistaken attempt to control their problems. Other people think that, in order to develop mental health problems, a person must have a genetic tendency to them; and that cannabis may act as a switch, turning on an underlying problem. This area needs much more research.

Withdrawal from cannabis

With some recreational drugs, users experience unpleasant withdrawal symptoms when the drug is no longer affecting their body. They feel that they must keep taking the drug in order to avoid the withdrawal symptoms. In this way, they become physically dependent on, or addicted to, the drug. The withdrawal symptoms make the drug very hard to give up. By contrast, for low-level, short-term users of cannabis, there are no physical effects when they stop taking it. Withdrawing from heavy, long-term cannabis use can

'People who smoke cannabis ought to be aware that it has equal effects to cigarettes on the body, and worse effects on the mind.' (Dr John Henry, St Mary's Hospital, London)

'Mum says a friend of hers tried cannabis in his teens, and soon after he started behaving strangely. They said he was schizophrenic. I don't think the drugs caused it, but I'm not sure. I don't want to take the risk.' (Miles, aged 14)

cause mild discomfort, but this is short-lived. Some reports suggest that long-term users of cannabis may have problems sleeping when they stop taking the drug. They may feel restless, irritable and anxious. With support and encouragement, these symptoms soon disappear.

It might therefore seem that it would be quite easy to stop taking cannabis. However, psychological dependence can be very powerful: the person may feel that they cannot get through their normal daily life without cannabis, and therefore it is very hard to 'kick the habit'. People who are psychologically dependent on cannabis may need support to help them find ways of relaxing or coping without the drug. This can be found through the family doctor or from some of the drug charities listed on page 62.

Talking down

An overdose of cannabis can cause extreme anxiety. The way to help a person in this situation is to talk to them calmly and reassuringly until the feeling wears off.

Overdose

Drug overdoses can harm the mind and body, and some can be fatal. It is not thought to be possible to overdose on cannabis. However, small children who take cannabis by mistake (sometimes in the form of a cake or cookie) can end up unconscious. This increases the risk of death, from inhaling vomit while unconscious.

Because the production of cannabis is not controlled, the strength of the drug varies. So, even though users think that they are taking the same amount of drug as normal, they can easily take a stronger dose. This can result in a fast heartbeat and feelings of extreme anxiety and fear. The best treatment for this is calm reassurance – sometimes called 'talking down', while the person waits for their body to break down the chemicals that are causing the sensation.

Sleepiness

Early in the twentieth century, anti-cannabis campaigners said that cannabis made people aggressive and contributed to violent crime. In fact, the opposite appears to be true. Cannabis tends to produce tranquillity and sleepiness, which are far removed from the activity required to be aggressive.

The sleepiness caused by cannabis definitely affects the user's ability to operate machinery safely and to drive. Cannabis reduces the ability to concentrate and to control fine movements and make fine adjustments. It also affects the perception of time and space. It is clearly very dangerous for someone to drive or to operate other machinery when they have taken cannabis. As the drug's effects can recur for up to three weeks, it is unwise to drive during the whole of that time.

A menace on the roads

'Cannabis is a menace on the roads. It disrupts drivers' ability to finely control movement and they become dangerous. At present we can't test for cannabis intoxication at the roadside, but we are working on tests. If we suspect that someone is under the influence of a narcotic substance, then we take them to the police station for further tests. The penalties for being intoxicated whilst in control of a car are severe, with good reason. When you are driving, you are in charge of a powerful machine that is capable of killing innocent people if it is not handled carefully. Imagine how you would feel if someone you love was killed by a car driver who was high. People have to take responsibility for their actions. You should never drive if you have taken anything that may affect your judgement, concentration or your ability to control a vehicle.'
(Alex, aged 24, traffic police officer)

Cannabis and addiction

Whether cannabis is addictive or not is a big health issue. For a drug to be considered addictive, one or more of the following must be true.

- The drug has to affect the user in such a way that they continue to use it regularly and repetitively, despite any unpleasant effects it may have on their bodies or their lives.
- The user has to experience a physical need or a psychological need to use the drug, and the need has to be so strong that they feel compelled to keep on using the drug.
- The drug has to activate the parts of the brain that are activated by normal pleasures, such as food, sex or laughing.
- Usually, the drug has to cause unpleasant symptoms for the user when they stop using it.

Not everyone who uses an addictive drug becomes addicted to it. It depends on many factors, such as their genetic makeup, family history, personality, mental health and general health.

It seems that it is possible to become psychologically dependent on cannabis. Some users feel that they come to rely on cannabis to help them cope with life. However, this is nothing like addiction to heroin and cocaine. These drugs take over the brain's ability to experience pleasure and are very powerful. Cannabis does not affect the pleasure areas of the brain and does not seem to produce a physical dependence.

'My sister tried cannabis at college. Mum was really worried that she would turn into a drug addict and drop out of college. She didn't though. I think she only tried it once or twice. Just wanted to make mum mad.'
(Karen, aged 17)

Users of heroin and cocaine require more and more of the drug to experience the same effects. This is called tolerance. The same can happen for cannabis users, but the increase in dose needed to produce an effect is very low and so tolerance to cannabis takes a long time to develop.

3 The cannabis plant today
Providing for modern needs

As we saw in Chapter 1, the cultivation of cannabis was generally banned in the twentieth century, in an effort to prevent people from using the plant as a drug. However, the arguments have always been put forward that the plant can be put to many uses and that it is valuable from an environmental point of view. Cannabis products are renewable (you can grow as much as you need) and biodegradable (they break down harmlessly when disposed of). Over the last part of the twentieth century, the laws preventing people from growing cannabis for hemp were gradually relaxed. It is now grown as a legal crop in many countries, for its fibre, seeds and oil.

Cannabis fibre – strong and durable materials

The most useful fibre produced from cannabis is hemp. It comes from the *cannabis sativa* variety of the plant, which grows tall and therefore produces long fibres. Industrial hemp production is now legal across most of Western Europe. Cannabis is also grown as an industrial crop in Canada and in four states in the USA: Hawaii, Minnesota, Vermont and North Dakota. However, it is uncertain whether the US Drugs Enforcement Agency will allow farmers to continue planting the crop. They fear that any relaxation of the laws on growing cannabis will increase the amount used as a drug. They feel that the risk of increasing the use of cannabis as a drug is much greater than the environmental and industrial benefits of using cannabis for fibre.

Traditional use
Being strong and durable, hemp rope is still used at sea, even though ropes made from new synthetic materials are also available.

Tough togs
Clothing made from hemp fibre has a reputation for being hard-wearing.

Hemp fibre is strong, hard-wearing and resistant to rotting. It has always been considered the best fibre for making rope and sacking, especially for use in ships and industry. Hemp fibre is also used to make cheap, hard-wearing clothing. Cloth made from hemp is very similar to cotton or linen. It is warm in winter and cool in summer.

Hemp fibre has recently found a new niche in the building industry. It is manufactured into 'press board' or 'composite board', which is much more durable and elastic than fibreboard made from wood pulp. French scientists have also developed hemp into a material that is used as a filler to insulate the walls of buildings, in order to prevent heat loss and reduce noise pollution.

'I have a pencil case made from hemp. I didn't realize that hemp was a form of cannabis.'
(Rachael, aged 12)

Another use of hemp fibre is in making string, paper, packaging, and plastics for items like skateboards and car bodies. The German company Daimler was one of the first to experiment with using hemp fibres in the construction of its cars, although they did not proceed because of legal problems with obtaining the fibre. Hemp is now used in the interiors of many Mercedes-Benz cars, and Daimler are looking at using hemp components in car bodies as well.

Cannabis fuel

Cannabis stalks contain lots of fibre and cellulose and these can be converted into fuel in two ways. One process, called pyrolysis, converts the stalks into a substance like charcoal, which can be burnt to produce energy. The other process involves fermenting the stalks to produce the chemicals methanol and ethanol, which can be used as a fuel. Hemp seed oil can also be processed into a form of petrol.

Cannabis seeds

Cannabis seeds are commonly used as an animal feed and are popular with birds. They are highly nutritious and, as the seeds from *cannabis sativa* contain hardly any of the intoxicating chemicals associated with cannabis, they are also considered by many to be a good food for humans.

Hemp seed contains approximately 25% protein, 30% carbohydrate and 15% fibre. It is a good source of essential fatty acids and it also contains carotene,

Diesel

The diesel engine was developed with the idea that it would be powered by fuel made from agricultural waste, such as hemp stalks. This did not happen, however, partly because of the laws against growing cannabis. The engines were converted to use petrol diesel instead.

Traditional recipes

Hemp seeds have a place in many countries' cooking. In Lithuania, hemp seeds are traditionally served with herrings. Raw or roasted hemp seeds are a traditional and popular snack in China. Hemp seed oil is used for cooking in the more remote areas of Nepal. In eastern Europe and the Baltic states, hemp was added to soups and stews and traditionally ground into a paste similar to peanut butter, which was eaten with bread. Before it was possible to separate hemp seed from its tough outer casing, this paste was hard and gritty. It is still used in eastern Europe and Russia and in places where it can be bought in food stores.

phosphorous, potassium, magnesium, sulphur, calcium, iron, zinc, and vitamins E, C, B1, B2, B3 and B6. It can be used as an ingredient in biscuits, snacks, veggie burgers and porridge. Hemp seed oil is not suitable for frying, as it does not store well, but it is used in a number of processed foods such as sauces.

Hemp oil

Hemp oil is reported to be good for the skin and so it is used in many cosmetic products such as hand cream, soap, shampoos and moisturizing cream. Hemp oil is also used as a lamp oil, in the printing industry, for lubricating industrial machines, in household detergents, and in a similar way to linseed oil in paints, varnishes, resins and stain removers.

'We bought some hemp hand cream for my gran. She loves gardening and it's supposed to be good for your skin.'
(Osma, aged 10)

Medical uses of cannabis

The properties of cannabis as a medicine have been recognized throughout history, but twentieth-century laws prevented doctors from using it to treat their patients. Today, cannabis cannot be prescribed by doctors. However, synthetic forms of cannabis have recently come on to the market, to treat some illnesses, under tightly controlled conditions. The laws governing what can and cannot be prescribed to patients are constantly under review, and so this situation may change in the future.

Improving the appetite

In Chapter 2 we saw that cannabis users talk about 'the munchies', to describe the hunger and increased appreciation of food that cannabis gives them. In the past, doctors used cannabis to treat diseases that reduce the appetite and cause profound weight loss. Diseases with that effect today include AIDS and some cancers, and some modern doctors would

Therapeutic or recreational

Therapeutic drugs are drugs prescribed by a medically qualified practitioner, or sold by a pharmacy, in order to cure or relieve the symptoms of an illness or injury.

Recreational drugs are taken by an individual in order to change their mood or their state of mind.

like to be permitted to prescribe cannabis for people with these conditions. They say that cannabis may help sufferers to feel hungry and to enjoy a meal – something that they find very difficult without assistance. Eating is one of our basic needs, and being unable to enjoy food can result not only in weakness and loss of weight but also in depression.

'I have leukaemia and the chemotherapy makes me feel very sick. I hate it. Every time I see the hospital now, I throw up. My doctor hasn't found a drug to stop me feeling so sick yet.'
(Boris, aged 14)

Reducing nausea

As well as stimulating the appetite, cannabis can reduce nausea (feeling sick). Nausea can be a problem for people undergoing chemotherapy treatment for cancer. Some doctors would like cannabis to be legalized for medical use, in order to help these patients. In the USA, a synthetic form of cannabis called Dronabinol was licensed for use as an anti-emetic (anti-sickness medicine) in 1985. In the UK, a synthetic equivalent of one of the chemicals in cannabis was licensed in 1982, as long as it was only used for patients for whom nothing else worked, and as long as it did not leave the hospital. Some patients feel that this is unfair, as many of them feel sick before, during and after hospital chemotherapy.

Chemotherapy
The drugs used in chemotherapy are very powerful, in order to control or destroy the cancer cells. The treatment can make the patient feel very sick.

Treating glaucoma

Glaucoma is a condition that affects the eyes. Your eyes are filled with a clear fluid, which helps to keep the eyes' round shape. Small channels allow excess fluid to drain away, so that the pressure within the eye remains steady. In glaucoma, the channels become blocked and the pressure gradually increases, and this affects the eyesight. Eventually the person can become blind. Glaucoma is one of the commonest forms of blindness and affects approximately 1.5% of 50 year-olds and 5% of 70 year-olds.

Treatment for glaucoma is difficult and has unpleasant side effects. Some studies of the effects of smoking cannabis have found that the drug reduces the pressure within the eye, although it is not certain how this works. Pharmaceutical companies are trying to develop an eye-drop that contains cannabis for treating glaucoma.

Taking cannabis for multiple sclerosis

Multiple sclerosis (often known as MS) is a medical condition that disrupts the normal function of the nerves, brain and spinal cord. The illness comes and goes unpredictably, and attacks can last for weeks or months. Each attack leaves the sufferer with slightly more disability. Because the spinal cord and brain are involved, the attacks can cause symptoms anywhere in the body. These can include tingling, numbness, blurred vision, tiredness, paralysis, muscle cramps, spasms and pain, loss of bladder or bowel control, constipation and depression. The causes of MS are not yet clearly understood, but it is probably linked to a problem with the immune system.

The muscle spasms, cramps and pain can be hard to live with and they are difficult to treat. Many of the medicines available have unpleasant side effects and can be addictive if used long-term. Cannabis has long been known for its effectiveness at reducing muscle pain, spasms and cramps. Today, there are many MS sufferers who risk prosecution by taking cannabis, because they feel it is the only drug that really relieves their pain.

Some people with MS report that, when they take cannabis, they regain bladder control. Some who are wheelchair-bound claim that cannabis relieves their symptoms so much that they can walk unaided. Individual, personal accounts like these, about the way a chemical affects a disease, are called anecdotal evidence and are not considered reliable by scientists. The World Health Organization feels that there is no conclusive evidence that cannabis does help people with MS. Properly set-up scientific studies are needed to work out whether cannabis is really an effective treatment.

As cannabis was listed in the USA as a schedule 1 controlled drug (a drug with no medical use) and was similarly controlled in most other Western countries, scientific trials were impossible. However, in the UK, a House of Lords committee discussed the issues surrounding the use of cannabis by people with MS and permission was given for trials to take place. The first results of these, published in 2003, seemed to show that cannabis can help relieve some of the pain and discomfort associated with MS, but does not cure the condition. More research is needed.

A difficult dilemma

'MS is a terrible condition. It comes and goes, and each time it flares up, it takes a bit more of you away. I spend most of my time in a wheelchair now. I'm too weak to walk far. I get terrible cramps in my legs. On the internet, some people with MS say that smoking cannabis has really helped them. I'm not sure. It's illegal and I don't want to go to prison. Maybe I'll try it if it gets worse, but then I don't know how to get cannabis or how to take it. The trouble is, there is so little treatment for MS. Everyone with it is desperate for a treatment that works.'
(Colin, aged 45, ex printer)

Cannabis for pain control

Cannabis has been used for centuries to treat muscular and nerve pain and there is some evidence to support its use as a pain reliever, particularly for chronic pain. Pain caused by cancer, for which traditional medicines cause nausea or other unpleasant side effects, may also be relieved by cannabis or synthetic cannabis-based medicines.

Before cannabis was declared illegal, it was used to treat menstrual cramps and also to relieve labour pain – the pain a woman experiences in childbirth. Because the drug is illegal, there has been little scientific study of the effect of cannabis on this form of pain. One main concern is how the drug might affect the baby before it is born.

'My mum gets really bad migraines. Sometimes she is ill for days with it. She'd never try cannabis though. She hates illegal drugs.'
(Roberto, aged 15)

Could cannabis be a treatment for migraine?

Migraine is a severe headache, which can occur with feelings of sickness, vomiting and blurred vision and can last for hours or days. It is usually brought on by a trigger, such as bright lights, eating certain foods, or hormone changes (for example, before a woman has her period). About 20% of adults have had a migraine at some point in their lives and many suffer from them frequently. In the nineteenth century, cannabis was a favourite treatment for migraine.

There has been very little scientific research into the effects of cannabis on migraine. Some doctors and pharmaceutical companies feel that it would be worth carrying out such research, to see if cannabis could be an effective treatment.

Migraine misery
One in five adults suffers from migraine.

4 Cannabis and the law
A 'hot' topic

In US law, cannabis is classified as a schedule 1 narcotic. Other drugs in this group include LSD, heroin and cocaine. Schedule 1 drugs are classified as such because they are highly addictive, of no medical use and their use is associated with crime and violence. The courts can impose the longest prison sentences for people found to be using, selling or smuggling these drugs. The United Nations Treaty 406 Single Convention on Narcotic Drugs, agreed in 1961, aimed to outlaw cannabis production and use around the world by 1991. Although this target was not achieved, the production of cannabis drug is now illegal in most countries. In the UK, in 1971, cannabis was classified as a Class B schedule 1 drug, under the Misuse of Drugs Act. This meant that anyone found in possession of the drug could face a long prison sentence. In 2003, cannabis was reclassified as Class C schedule 1. This means that it is still illegal to possess cannabis, but people caught only once with a very small amount will not be prosecuted. People found with more than a very small amount will be prosecuted as dealers.

Ever since cannabis was made illegal, there has been protest. Some doctors have rejected the idea that cannabis has no medical value. Users have argued that it is not a dangerous drug. Manufacturers have wanted to use hemp

British drugs haul
Large quantities of cannabis are illegally trafficked every year. This haul of 18 tonnes was discovered at a warehouse in Kent in 1996.

Gateway drugs

People who become addicted to drugs such as heroin or crack cocaine have nearly always used another more socially acceptable, 'soft' drug before they were introduced to heroin or cocaine. Common examples of these soft drugs include alcohol, tobacco, cannabis, steroids and inhalants (glue sniffing). The theory is that these drugs act as a 'gateway' to 'hard' drugs. Using a gateway drug does not directly cause someone to go on to try other drugs, but it does introduce them to people who have access to these types of drug. It also introduces them to the behaviour associated with drug use, such as deceiving other people about using drugs, and using drugs at specific times. Therefore they are more likely to try another drug than someone who does not use soft drugs.

fibre in industry. There have been continuing debates all over the world about whether cannabis should be 'declassified'. Other people continue to view cannabis as a highly dangerous 'gateway' drug.

Cannabis crimes
⊛ Growing cannabis

It is a crime in most countries to grow cannabis or to allow your home to be used for growing cannabis. In many countries (including France, Spain and the UK), a licence can be obtained from the government to grow the varieties of cannabis that have very low concentrations of psychoactive chemicals, for fibre, seed and oil use.

⊛ Processing cannabis

It is a crime to process cannabis. This includes harvesting the drug-laden resin from the flower tops or collecting and drying the leaves and seeds, whether it is for personal use or to be sold on to other people.

⊛ Smuggling and trafficking

It is a crime to import or export cannabis, whether you intend to supply other people or use the drug yourself. People convicted of trafficking drugs tend to receive long prison sentences, as these are considered the most serious crimes.

Supplying and dealing in cannabis

It is a crime to supply cannabis to another person, whether you are paid for it or not. People convicted of supplying cannabis also tend to get long prison sentences. Currently in the UK it is possible to be sentenced to 14 years in prison for possession of cannabis with intent to supply others.

Possessing cannabis

The possession of cannabis is a crime in most countries, but the penalties in different countries range from cautions to fines to long prison sentences. In the UK, until cannabis was reclassified as Class C, someone found guilty of possessing cannabis might be sentenced to up to five years in prison plus a fine. Since the reclassification, possession is no longer an arrestable offence unless the person is in a sensitive place such as outside a school or playground. People found with a very small amount of cannabis for personal use will be cautioned. If repeatedly found in possession of the drug they may be prosecuted and liable to a fine or up to two years in prison. Anyone found with more than a small amount will be prosecuted as a dealer.

Illegal crop

People who grow cannabis illegally often cultivate the plants in cellars or basements, under artificial light and heat.

Cannabis and coffee

Customers in certain cafés in Amsterdam are permitted to use small amounts of cannabis.

Different approaches

Some countries have experimented with different ways of managing cannabis use. In 1974, the Netherlands changed its drugs laws so that some coffee shops are licensed to sell small amounts of cannabis, under tightly controlled conditions, for use only on their premises. Some people see this project as a success. Recent statistics show that the percentage of young people using cannabis in the Netherlands is smaller than in several other European countries. Other people feel that the availability of cannabis has helped to make the Netherlands a haven for drug smugglers and drug users, although there does not seem to be much evidence to support this opinion.

The laws on the commercial production of cannabis were relaxed in the UK and across much of Europe in the 1980s and a few farmers started small-scale production. Hemp products and cosmetics and foods containing hemp oil are

Statistics

Percentages of people aged 15-34 who had used cannabis in the last 12 months:

Denmark	13.1%
Finland	4.9%
France	17%
Germany	13%
Ireland	17.7%
Netherlands	9.8%
Norway	8.1%
Portugal	6.2%
Spain	12.7%
Sweden	1%
United Kingdom	16.6%

(from a 2002 report)

commonly on sale. In the UK there have been experiments in relaxing the laws for possession of small amounts of cannabis for personal use. It was felt that police were spending too much time arresting and processing people who were found in possession of very small amounts of cannabis for personal use. In fact, over two-thirds of all drug-related offences in the UK consisted of the possession of small quantities of cannabis. It was argued that the police time spent on these could be put to better use targeting the people who organize the illegal drugs trade.

In a pilot study in Brixton (south London), people found with small amounts of cannabis were cautioned rather than being arrested. The study was generally thought to have been successful in re-targeting police time to deal with crimes involving drugs like crack cocaine and heroin. As a result of this and of a general relaxing of public attitudes in the UK, cannabis was downgraded from a Class B to a Class C drug, with effect from January 2004. This meant that cannabis remained an illegal and controlled drug, but was in the same category as steroids rather than being classified with heroin and cocaine. Possessing cannabis would no longer be an arrestable offence, but people could still be prosecuted for possession. The police might adopt a 'three strikes and you're out' policy: anyone found in possession of cannabis three times in a year would be prosecuted. Anyone found with more than a very small quantity of cannabis would be considered a dealer and prosecuted as such.

The situation is very different in the USA. During the late twentieth century, several states relaxed their laws on the medical use of cannabis. People with chronic pain or medical conditions that might benefit from treatment with cannabis were therefore able to use the drug without fear of going to prison. There were also attempts by some states to restart the growing of cannabis for hemp fibre. However, since 2000, the Drugs Enforcement Agency has clamped down and is trying to stop the sale of any cannabis-based food or cosmetic product.

Arguments for legalizing cannabis use

The following are some of the points used by people who argue that cannabis use should be legalized.

- Cannabis is a naturally occurring herb with beneficial properties.

- Cannabis is not physically addictive.

- Cannabis provides a possible medical treatment for several symptoms, diseases and conditions that are hard to treat using normal medicines. These include muscle spasm in multiple sclerosis, nausea connected with chemotherapy, the wasting associated with AIDS, and the eye condition called glaucoma. People who have painful and disabling diseases and who find cannabis helpful have to commit a crime in order to obtain cannabis for medical use. Many people argue that it is absurd to threaten prison sentences to very sick people who buy cannabis in a desperate attempt to find relief from pain or the wasting associated with some forms of cancer and AIDS.

- Cannabis has been used for thousands of years and the plant has many other practical uses, including for making rope, cloth, paper, fuel for cars, biodegradable plastics, animal feed, cosmetics, detergents and building materials.

- It is preferable to use cannabis for these products, as they are less damaging to the environment than petroleum-based products. Cannabis is a renewable crop, which does not need large amounts of pesticides to grow healthily. Also, many of the products made from cannabis are biodegradable, which means that, when they are thrown away, they do not pollute the environment.

- It is estimated that, in the near future, nearly a third of all adults in the UK and France will have tried cannabis at some point during their lives. A very small percentage of these people go on to use cannabis on a

'Cannabis is not like the other drugs, heroin or cocaine. It's never killed anyone. I think we should be allowed to make up our own minds. If you're old enough to smoke tobacco, why shouldn't you be allowed to smoke cannabis?'
(Elsie, aged 43, florist)

regular basis, and an even smaller number go on to try drugs such as heroin or cocaine.

- The police time spent on arresting or prosecuting people for possession of cannabis could be used more profitably, for tracking down the people who profit the most from illegal drug use and who are responsible for selling drugs to the most vulnerable.

- Cannabis is attractive to young users because it is illegal. Controlling the use and sale of cannabis and legalizing it will make using cannabis a less attractive pastime.

- In the Netherlands where the use of cannabis is legal, in well-regulated environments, only 0.2 – 0.5% of those who try cannabis go on to become problem drug users

- Cannabis is much less harmful and much less addictive than many legal drugs, including alcohol and tobacco.

'I think we should downgrade cannabis. So much of our time is spent dealing with people found in possession of a small amount for personal use. It's time that could be better spent going after the big boys.' (Simon, aged 28, police officer)

Campaigning
In many countries people have such strong opinions about cannabis that they take part in protests to try to change the law about the drug.

Arguments against legalizing cannabis use

People who argue that cannabis use should remain illegal make the following points:

- Not all naturally occurring herbs are beneficial. For example, tobacco and deadly nightshade are herbs but they are toxic. Whilst cannabis is not as toxic as these, it does have some negative effects on health. Smoking cannabis mixed with tobacco puts you at risk of developing lung, throat and mouth cancers and there is some evidence that taking cannabis puts you at risk of developing mental health problems and heart disease.

- Cannabis users can become psychologically dependent on the drug.

- The medical uses of cannabis have not yet been fully researched, so officially, the jury is still out on whether or not cannabis is a safe and reliable treatment.

- The fact that a drug is a medicine does not mean that it is safe for everyone to use. For example, heroin is used medically to treat severe pain, but it is highly addictive.

'I am strongly against the legalization of cannabis. As far as I'm concerned, there is no such thing as a harmless drug. They all lead people into a life of crime and misery.' (Stella, aged 24, police officer)

- There are two main species of cannabis plant. *Cannabis sativa* grows very tall, producing long fibres suitable for cloth, rope and papermaking. It is farmed in many parts of the world and this is acceptable, since it contains only low concentrations of the chemicals that make cannabis psychoactive. To obtain hemp fibre, there is no need to grow *cannabis indica*, the type of the plant that is used for its psychoactive properties.

Some people feel that it is unacceptable even to allow the cultivation of *cannabis sativa*. They feel that allowing products made from hemp fibre or oil onto the general market gives people the idea that cannabis is acceptable; and they argue that, therefore, all cannabis products should be banned.

- It is misleading to talk about cannabis as a 'soft' drug, which is less harmful than 'hard' drugs like heroin and cocaine. People making this point argue that there is no real difference between hard and soft drugs; every intoxicating drug has the potential to cause harm, both to the individual using it and to society as a whole.

- If cannabis is legalized, the people who used to try it because it was risky and illegal will move on to the more dangerous drugs that are still illegal, in order still to be daring.

- The United Nations has criticized the UK's decision to reclassify cannabis. They feel that this sends the wrong message to people growing the crop in other countries and puts at risk the worldwide effort to reduce the production of illegal drugs.

- Legalizing cannabis would probably cause a big rise in its use and therefore a big increase in the amount of money spent on health care to deal with diseases associated with smoking it.

'I don't think they should legalize cannabis. I see drug addicts coming in here every day, up to no good, thieving. They need to be locked up, not left on the street.' (Jim, aged 53, newsagent)

The arguments about cannabis continue around the world. In the next chapter we investigate who uses cannabis and its impact on today's society.

Road to addiction

Some people argue that people who take cannabis are more likely than others to go on to try more dangerous drugs such as heroin.

5 Cannabis in today's society
Considering your viewpoint

Cannabis is used by such a large and varied section of society that it is impossible to characterize a typical cannabis user. Many people first come across cannabis when they are in their late teens or early twenties. It is estimated that, in some parts of the world (notably the UK, USA, France and Australia), well over 20% of adults will try cannabis at some point. Of these, only a small fraction will continue to use cannabis regularly, and it is estimated that less than 1% go on to try more dangerous drugs such as heroin.

Since the Second World War cannabis use has grown consistently, whereas other drugs have gone in and out of fashion. There are several reasons for the increase in cannabis use. People have become richer and so have more money to spend on non-essential things. Many people see cannabis as a less risky illegal drug than others. Thirdly, cannabis is increasingly available.

All types
Statistically, one of these five teenagers will try cannabis at some time in their life.

'I don't want to take cannabis. I like to stay in control of my body.' (Saskia, aged 15)

The power of public opinion

Many countries around the world are introducing legislation to decriminalize the possession of small amounts of cannabis. One of the reasons for this change is that large sections of the public are in favour of such a move. A recent survey carried out in Britain, about public attitudes to cannabis, found that 54% of the adults questioned thought that cannabis should be legalized and 86% thought that doctors should be able to prescribe it.

Sweden has been one of the most successful European countries at limiting the rise in cannabis use. The European Union estimates that under 8% of adults in Sweden have tried cannabis. Sweden has harsh legal penalties for the possession of cannabis and, unlike in other countries, a person in Sweden can be prosecuted if the police think that they may have taken cannabis, even if no drugs are found on the person. However, the success of the anti-cannabis campaign in Sweden is not only due to its penalties. Sweden has a long history of public intolerance to the use of any intoxicating substance, including alcohol and tobacco. Swedish public opinion is very much against the use of any recreational drug, legal or otherwise, and so very few people there want to buy cannabis.

Other people suggest that Sweden's success is because it borders countries that have a more tolerant approach to recreational drugs – so that anyone wishing to get drunk or 'stoned' (high on cannabis) crosses the border into Denmark or Norway.

Drug use

Countries around the world have combined to try to reduce illegal drug use and drug crime, and – if you ignore the growing use of cannabis – their efforts have been fairly successful. The numbers of people taking drugs that are obviously very risky, like cocaine and heroin, are stable (not going up) or reducing.

'In Sweden we don't like drugs. They are not good for the body or the mind.'
(Goran, aged 16)

Cannabis and the environment

The campaign for the legalization of cannabis is often associated with 'environmentalists'. They emphasize the need to prevent pollution and waste and to ensure that planet Earth continues to be able to sustain life. Environmental pressure groups support the legalization of cannabis mainly because the cannabis plant has multiple uses. The campaigners point out that growing cannabis for hemp fibre and oil would help solve some of today's environmental problems. For example, using hemp fibre to make fibreboard, building materials and even car bodies has an important advantage: cannabis is considered a 'carbon dioxide neutral' plant.

Carbon dioxide is produced by most industrial processes that involve burning fuel. It is one of the pollutants known as greenhouse gases, which are thought to build up in the atmosphere and contribute to global warming. The processes needed to turn hemp fibre into fibreboard, plastic or paint produce carbon dioxide; but this is

Air pollution
Waste gases from industry pollute the air, whether the industry uses petrochemicals or hemp fibre. But growing more hemp to use in industry could help reduce air pollution.

Global warming and greenhouse gases

Global warming is a term used by scientists to refer to increases in temperature on the Earth, caused by the effect of 'greenhouse gases'. Global warming may lead to changes in rainfall pattern around the world, a rise in sea levels, and a change to the climate that would affect plant and animal life.

Greenhouse gases are gases present in the Earth's atmosphere, which trap heat from the Sun. Without these gases (which include water vapour, carbon dioxide, nitrous oxide and methane), the Earth would be very cold. The concern is that human activity is producing so much of these gases that we are in danger of making the Earth warmer than it would naturally be. Even a small rise in the average temperature of the Earth can cause problems for humans, animals and plants.

balanced (or 'neutralized') by the fact that, while the cannabis plants are growing, they absorb lots of carbon dioxide from the air and give out oxygen, through photosynthesis. Taking this into account, using hemp fibre in industrial processes is said to produce less carbon dioxide than using petrochemicals. Growing more cannabis, in order to use hemp fibre rather than petrochemicals in more industrial processes, would therefore reduce the greenhouse gases in the atmosphere.

Cannabis fibre is considered 'environmentally friendly' in other ways too. For example, cloth made from hemp looks and feels very similar to cotton, but growing a crop of cannabis requires far fewer pesticides than growing cotton plants. Therefore, people argue that using hemp fibre rather than cotton for making cloth would reduce the usage of poisonous chemical pesticides, which are dangerous to animals and humans and which pollute the environment. Also, cannabis plants can be grown at a much higher density than cotton plants, and this means that more fibre can be produced on less land. This, as well as the smaller amount of pesticides needed, makes hemp fibre cheaper to produce than cotton.

What is your attitude?

It is likely that some people you know may use cannabis and that you may be invited to try it. It is a good idea to prepare yourself in advance for situations where you may encounter cannabis. If someone offered you a cannabis cigarette, what would you do? If you want to try it, are you fully aware of the risks you are taking? Do you know what it is you are being offered? What is the best way to say 'no'? Practising saying no in different situations is called role-play. It is usually better if you can role-play with a friend, so you can discuss different ways of handling a situation. As you think about your attitude to cannabis as a drug, you need to be fully aware of the risks.

The risks involved in taking cannabis

⊛ You can never be sure exactly what you are taking. Some drug dealers mix or 'bulk up' the drugs they sell with other, less valuable substances, in order to increase their profit. Although cannabis is less easy than other drugs to adulterate in this way, the possibility is still there. The unknown substances mixed with the cannabis could be dirty and dangerous to your body.

⊛ You do not know the concentration of chemicals in cannabis, so you can never be sure of what the effects will be. High-grade cannabis can cause severe anxiety and can be very unpleasant.

⊛ You may cause an accident if you try to drive a car or operate machinery under the influence of cannabis.

What would you do?

If you found yourself in a situation where people were using cannabis, what would you do? Knowing about the drug and all its risks will help you make your decision with confidence.

Expelled

'My brother was expelled from school. There were undercover cops at a nightclub he was at and they caught him buying some pot from a dealer they were watching. It was really bad. Mum and Dad went ballistic. Mum wouldn't stop crying, saying that he had ruined the family's good name. He was about to take his exams too. He was just trying to show off in front of his girlfriend – but she wasn't impressed when he was arrested. He got a caution from the police. Because he was expelled, Mum and Dad had to find him a private tutor. I think he's going to be grounded for the rest of his life. It was awful.'
(Sophie, aged 14)

◉ Mixing drugs, or taking a drug when you are already under the influence of another, is very dangerous. It is especially dangerous to take any drug at the same time as you are drinking alcohol or taking cold medicines.

◉ There is a real risk of being caught by the police. You may be arrested and prosecuted. This can have a very bad effect on your future job prospects.

◉ You run the risk of developing lung disease and possibly mental illness if you take cannabis regularly.

'There's too many things to do to waste time taking drugs. I want to be in there making things happen, not out of my head.'
(Cameron, aged 17)

◉ There is a real risk to other members of your family. Your parents may be arrested and charged if drugs are found in their home.

◉ You risk being expelled or suspended from school.

◉ You run the risk of becoming psychologically dependent.

◉ Cannabis can make you so 'laid back' that you stop being bothered by anything. Although this may seem a positive effect, it can mean that you stop making the effort to maintain good relationships with the people who are important to you, and that you stop trying to succeed at school or work. Therefore it can have a very damaging effect on your future career and family life.

Glossary

addiction when someone uses a drug repetitively and compulsively, even though it has negative effects.

addictive causing addiction.

bhong a pipe used for smoking cannabis which contains water to cool the smoke before it enters the lungs.

chronic lasting a long time. Chronic pain is pain that lasts for more than a few months.

craving a desperate longing.

dependence physical dependence on a drug is a condition where, if the person suddenly stops taking the drug, their body reacts badly and they feel unwell. Psychological dependence is where the person feels a strong need to take the drug, which is not related to any physical changes in their body.

drug a chemical that is taken into the body in order to change the person's physical or mental state.

endorphins pain-relieving chemicals that are produced naturally in the brain.

fibres fine thread-like strands which can be of animal, plant or manmade origin. Many fibres are twisted or woven together to produce materials like sewing thread, rope, string or cloth.

flowering tops the flower parts of a plant.

glaucoma a condition where the pressure within the eye increases. This can result in sight defects and eventually blindness.

hallucinogenic causing the perception of something that is not really there (a hallucination). This can involve seeing, hearing, smelling, tasting or feeling.

hemp the name used for the fibres produced from the cannabis plant, which can be processed into materials such as cloth, rope or building materials. Sometimes people refer to the cannabis plant as a hemp plant and sometimes the word 'hemp' is used to refer to the less psychoactive variety of cannabis which is often grown for its fibre, oil and seeds.

high slang for the experience of being under the influence of a psychoactive drug.

hormones a group of chemical messengers produced by glands or organs in the body and which affect other parts of the body. For example, the hormone progesterone is produced during the female menstrual cycle and prepares a woman's body to become pregnant.

immune system the system in the body that protects it from infection.

inhale to breathe in. People who smoke cannabis inhale the smoke in order to become intoxicated.

insulate to prevent heat, cold or electricity passing from one side to another. People insulate their homes to prevent heat from escaping during the winter and so reduce their heating bills.

intoxicant something that makes you feel drunk, excited and/or elated.

intoxicating causing the experience of being drunk, excited and/or elated.

menstrual cramps painful cramps of the lower abdomen experienced by some women at the start of menstruation (periods).

munchies slang term for the hunger people feel after they have taken cannabis.

narcotic a drug that affects the brain to produce dizziness, euphoria, sleepiness and eventually unconciousness and death.

nausea the feeling that you may be going to vomit.

neuro-transmitters chemicals that transmit a message across a junction (synapse) between two nerves. Neurotransmitters are mainly found in the brain and spinal cord.

opiates drugs made from the opium poppy. These include morphine, codeine, opium and heroin.

overdose taking too much of a drug. This usually causes physical or mental damage.

pesticides chemicals used to kill insects that eat and destroy crops.

psychoactive affecting the brain and behaviour.

recreational leisure-time; recreational drugs are drugs used to achieve a 'high' rather than for any medical reason.

resin a sticky substance produced in the sap of some plants and trees.

selectively bred genetically engineered by selecting plants or animals that display a certain desirable quality and breeding only from them until a variety of plant or animal is produced which always exhibits the quality required.

smoking breathing in the chemicals produced by burning drugs.

stimulants drugs that increase the activity of the body and mind.

therapeutic to do with healing the body or mind.

tolerance the ability to endure something without showing serious effects. If the body is regularly exposed to a particular drug, it may become tolerant to it. This means that the body learns to react to minimize the effects of the drug and so the user needs to take an increased dose of the drug to get the same effects.

traffick illegal trading in a drug or smuggling a drug into a country.

withdrawal the process the body goes through when a person stops taking a drug.

Resources

Further reading

C. Kuhn, S. Swartzwelder and W. Wilson, *Buzzed, The straight facts about the most used and abused drugs from alcohol to ecstasy*, W. W. Norton & Company. New York, London, 1998.

Aidan Macfarlane and Ann McPherson, John Alstrop, *The new diary of a teenage health freak*, Oxford Paperbacks, 1996.
Funny, easy-to-read introduction to all health education issues including drugs.

Sarah Lennard-Brown, *Health Issues: Drugs*, Hodder Wayland, 2001.
The facts and issues surrounding illegal drug use.

Film

Withnail and I, 1987, directed by Thom Eberhardt. In this British film, set in the 1960s, two out-of-work actors try to deal with life through a haze of drugs. The film deals lightheartedly with the choice between real life and living in a drug-induced fog.

Organization

DrugsScope
Waterbridge House, 32-36 Loman Street, London SE1 0EE.
Telephone: 020 7928 1211
Website: www.drugscope.org.uk
An independent British charity which researches drug-related issues and advises on policy-making and drugs prevention and information issues. It has an excellent data base of information about all drugs.

Sources

The following sources were used in researching this book:

M. Bloor and F. Wood (editors), *Addictions and problem drug use*, Jessica Kingsley Publishers, London, 1998.

D. Emmett and G. Nice, *Understanding Drugs*, Jessica Kingsley Publishers, London, 1998.

S. G. Forman, *Coping skills and interventions for children and adolescents*, Jossey-Bass Publishers, San Francisco, 1993.

C. Kuhn, S. Swartzwelder and W. Wilson, *Buzzed, The straight facts about the most used and abused drugs from alcohol to ecstasy*, W. W. Norton & Company, New York, London, 1998.

A. Macfarlane and A. McPherson, *Teenagers, The Agony, the Ecstasy, the Answers*. Little Brown and Company, London, 1999.

Index

TiME TRAVEL GUIDES

ROMAN BRITAIN and Londinium

Ben Hubbard

FRANKLIN WATTS
LONDON • SYDNEY

Franklin Watts
First published in Great Britain in 2018
by The Watts Publishing Group

Artwork by:
Design: Collaborate Agency
Editor: Sarah Silver

ISBN 978 1 4451 5730 6

Printed in China

Franklin Watts
An imprint of
Hachette Children's Group
Part of The Watts Publishing Group
Carmelite House
50 Victoria Embankment
London EC4Y 0DZ

An Hachette UK Company
www.hachette.co.uk
www.franklinwatts.co.uk

CONTENTS

LONDINIUM

Welcome to Londinium – a former fortress town on the fringes of the Roman Empire which grew into the great cosmopolitan capital of Britain. Shown here in the 21st century, London is the English city where archaeologists regularly dig up objects and ruins that help to reveal who lived in Londinium about 2,000 years ago.

Time to go back

Now, however, we can use time-travel technology to journey back to Londinium in person. Strap in and get ready to sail up the River Thames to the bustling port, socialise at the public baths, and watch gladiators bash it out at the amphitheatre.

Your time travel guide

Congratulations for buying this Time Travel Guidebook – the must-have companion for travellers journeying into the past. The guidebook will give you expert advice on where to stay, what to see, and how best to spend your time in Londinium in the CE 2nd century. Top travel tips throughout will give you the low-down on what to bring, where to shop and which local foods to sample.

4

Beware of counterfeit coins

You'll need to exchange some currency as soon as you arrive, but look out for the counterfeit coins currently in circulation. Made mainly of bronze, these fake coins are much lighter than the official silver versions.

What to wear

Togas, tunics and sandals are all in fashion in Londinium. Romans often prefer theirs in the standard white, but wool cloaks with tartan patterns are also popular with the locals. If you fancy trying the latest styles, get ready to shop at the stalls and shops in the forum.

PUTTING LONDINIUM ON THE MAP

It's hard to believe that before the Romans arrived Londinium didn't exist. Instead, all that was here were a few farms scattered along the River Thames. Now it is a big, bustling city, which boasts a stunning array of must-see sites. Each of these sites has been highlighted on the map below. To learn more about these sites, simply turn to the page numbers given in the map key below.

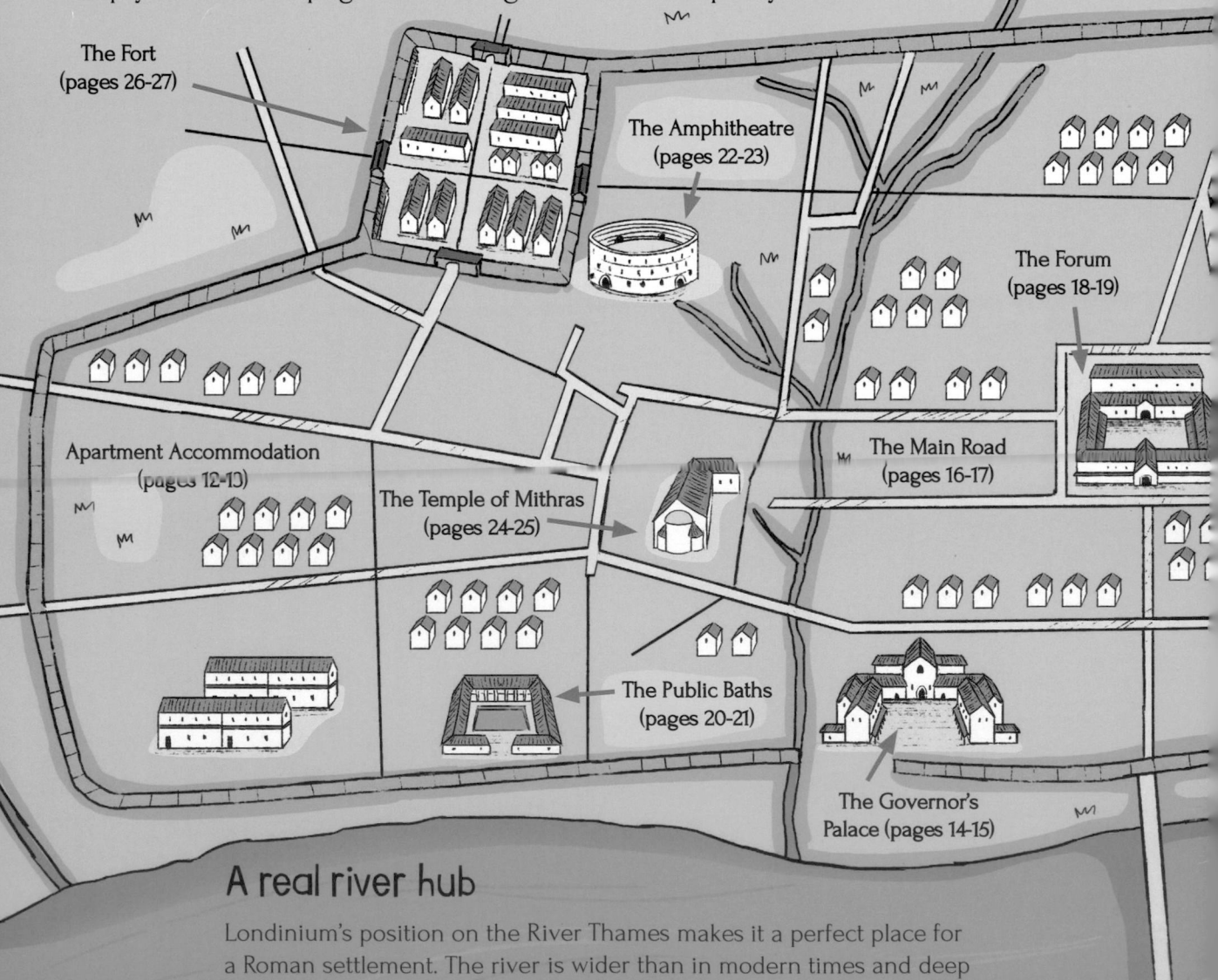

The Fort
(pages 26-27)

The Amphitheatre
(pages 22-23)

The Forum
(pages 18-19)

Apartment Accommodation
(pages 12-13)

The Temple of Mithras
(pages 24-25)

The Main Road
(pages 16-17)

The Public Baths
(pages 20-21)

The Governor's
Palace (pages 14-15)

A real river hub

Londinium's position on the River Thames makes it a perfect place for a Roman settlement. The river is wider than in modern times and deep enough for merchant sailing ships to deliver their goods at the port. Its position on the river is why it became a great trading centre.

Learn some Latin

People come from all over the known world to trade in Londinium and you'll hear different languages wherever you go. The language spoken by the Romans however, is Latin.

Try learning these phrases (right), which may come in handy:

bene ('ben-ay') – Fine

mihi ignosce ('mee-hee ig-nos-kay') – Excuse me

vescere bracis meis! ('wes-kay-ray bra-kees may-ees') – Eat my shorts!

sentio aliquos togatos contra me conspirare ('sen-tee-oh alee-kwos togatos kontra may con-spee-rah-ray') – I think some people in togas are plotting against me

The Romans were famous for their roads which they used to connect and control their vast empire. This map shows the Roman road network of Britain.

The Port (pages 10-11)

Eburacum

Camelodunum

Lindum

Deva

Glevum

Isca

Noviomagus

Londinium

7

LONDINIUM: A HISTORY

The ancient Romans' long history with Britain began when the general, Julius Caesar, led two expeditions there in 55 and 54 BCE. However, after winning several battles against the tribal Britons, Caesar set sail again. Nearly a hundred years later, in CE 43, the Romans returned to conquer most of the country. They founded Londinium in around CE 50. It grew from a small settlement on the edge of the Empire into a thriving trading town and became the centre of government for Britannia, the Roman name for Britain.

Invasion with elephants

When Emperor Claudius ordered the invasion of Britain in CE 43 he wanted it to be fast and deadly. Over 40,000 Roman legionaries equipped with war elephants and siege machines easily defeated the tribes in England's south-east. Then, they built a bridge across the Thames so they could cross the river. This was the original London Bridge.

Meet the Britons

The native Britons are made up of tribes who now live under Roman rule. Some Britons hate the Romans and rebel against them. Others have moved to towns like Londinium and adopted the Roman way of life. However, the Britons have kept many of their own customs, clothing styles and gods. The Romans don't mind this, as long as everyone keeps the *Pax Romana* (Roman Peace).

Top Tip

Watch out for butchering Britons

In CE 60, Queen Boudica of the Iceni tribe was treated very badly by the Roman conquerors. In revenge, Boudica and her supporters attacked three Roman towns, including Londinium, massacring their inhabitants and burning the towns to the ground. Even now, attacks sometimes take place on Londinium – so be on your guard!

The Port

ARRIVE BY RIVER

A river cruise along the Thames is the best way to enter Londinium – even for modern visitors with time machines! This way you'll get a great view of the city as you sail under Londinium bridge and dock at the busy port. The port serves as an international trading hub for goods from around the Roman Empire.

Poke around the port

The port is your first stop and one of Londinium's most interesting sites. Here you'll see ships bring in glass from Germany, red pottery from North Africa, spices from the Middle East and amphorae (pottery jars) full of wine and olive oil from Greece, Spain and Italy. Being shipped out is the copper, silver, lead, iron and tin that attracted the Romans to Britain in the first place. Another famous British export is a waterproof woollen cloak called a *birrus Britannicus* – a must-have item for time-travellers visiting in winter.

Top Tip

Ancient adverts

Look out for adverts stamped on to amphorae and pots at the port. A recent popular advert is by a merchant selling his pottery flasks in Londinium. Each flask reads: 'Lucius Julius Senis's ointment for roughness of the eyes'.

Look up at Londinium bridge

Built from wood, Londinium's bridge is a few metres away from the 21st century London Bridge. Like the modern Tower Bridge it even has a special drawbridge that can be raised to let the masts of sailing ships through.

See the ships

Large, round-bottomed merchant ships from distant Mediterranean harbours and flat-bottomed barges for local transportation all dock at Londinium's port. Smaller boats are available for hire if you fancy a paddle upstream.

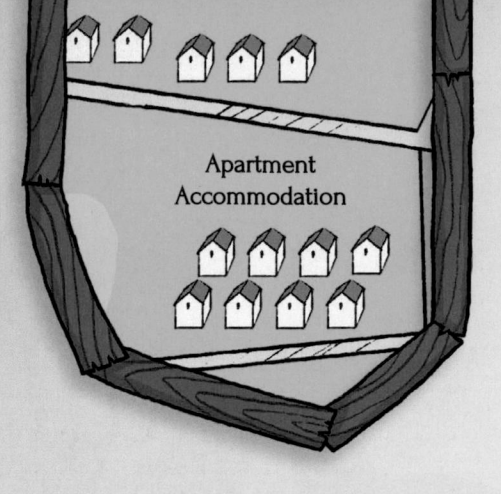

Apartment
Accommodation

WHERE TO STAY

To soak up city life at street level, book a single-storey apartment built along Londinium's main roads. These are simple homes for the labourers, shopkeepers and craftspeople working in the city. Don't expect any mod-cons (modern conveniences, like a washing machine or shower), but a basic bed, dresser and washing bowl are included in the price.

Your Roman home

Ordinary Londinium houses are built from timber and brick walls and are separated from each other by narrow alleyways. At the back are the living quarters and courtyard, with a shop or workshop at the front. Most residents keep animals such as chickens and pigs in their courtyards and have an outhouse for guests. This accommodation is perfect for time-travellers wanting to experience authentic, everyday Londinium.

Fill up on street food

Join the other Londinium residents and grab a snack or meal from one of the many street taverns and stalls. Here you can sit and enjoy a beef, pork or mutton stew from a steaming earthenware pot or grab a loaf of bread, baked fresh that day. Cucumbers, walnuts and peas are also popular local favourites.

Street life

The streets of Londinium can be an assault on the senses! You'll have to watch your step among the donkey dung piles and liquid waste babbling past in the roadside drains. There are smells of baking bread and simmering stews wafting from the roadside taverns. But drowning everything out are the city sounds, as cocks crow, pigs grunt and the shopkeepers' shouts mingle with the conversations of passing pedestrians.

The Governor's Palace

UPGRADE YOUR ACCOMMODATION

For time travellers with a few extra coins in their pockets, a stay at the governor's palace can sometimes be arranged. This will enable you to experience the luxurious living only enjoyed by the Roman aristocracy. To round off your stay, you will attend a gala dinner party thrown by the governor himself.

Modern comforts

Covering an area larger than two football fields, the governor's palace is an imposing building built to impress. Overlooking the river are large state rooms and a courtyard garden with an ornamental pool. Inside, lavish mosaics cover the floors and colourful frescos decorate the walls. Maximum comfort for guests is guaranteed with the latest Roman mod-cons: private baths and underfloor heating. Known as a hypocaust, slaves keep this underfloor system in operation by burning wood in the cellar furnace so that the warm air circulates under the floor.

Dinner party date

Get ready for a long evening of extravagance at the governor's dinner party. These events are designed to show off the host's wealth and power and no expense will be spared. You'll recline with the other guests on long couches, enjoy watching slaves dance and play music, and feast on several courses that will include eggs, seafood, meat, poultry and garum sauce. Fruit, nuts and honey cakes will follow, all washed down with Italian wine. However, be careful not to drink too much. Drunkenness at Roman dinner parties is considered distasteful.

Top Tip

Don't gag on the garum

No trip to Londinium would be complete without tasting garum – a salty sauce made from fermented fish guts, which is used like ketchup by the Romans. Called the 'noble sauce', garum is an expensive treat only enjoyed by the rich. Therefore, try not to turn your nose up at it!

The Main Road

GETTING AROUND

Like most Roman towns, Londinium has been constructed in a rectangular shape with a grid system of roads. Roman roads are flat, straight and easy to march an army along at speed. The Romans use their roads to keep control of their empire, but they are also a great way for time-travellers to hot-foot it across Londinium and check out the sites.

Top Tip

Try a luxurious litter?

Most Londinium residents make their way around on foot, or by horse, wagon, chariot or mule for longer journeys. However, if you fancy trying the deluxe option then why not hire a litter? A litter is a chair or couch attached to long poles, which is carried by slaves on their shoulders. Some litters also have a curtain around them so you can travel incognito, like a real Londinium celebrity.

Look at the legionaries

As you walk around Londinium, you may notice Roman legionaries building new roads. This is because a road is one of the first things the Roman army constructs when it conquers a new territory. Roman legionaries are therefore expert road-builders. Londinium's roads are constructed from layers of stone packed down to form a hard, slightly curved surface that drained rainwater into ditches.

Cursed luck

Londinium residents are typical city dwellers who can sometimes have a short temper. Beware of getting on their bad side, for you may become the subject of a curse. Often inscribed in pieces of lead, these curses are intended to cause the victim everlasting harm. A recent example from a jilted lover reads: 'I curse Tretia Maria and her life and mind and memory and liver and lungs'.

The Forum

SEE THE FORUM

As the political and social centre of the city, the forum is a must-see for visitors to Londinium. The forum is home to many shops and the basilica, or town hall, the largest building in Roman Britain. Best of all is the market in the uncovered central square. This is a great place to watch people buy food grown on British farms or brought by ship from other parts of the Empire.

Top Tip

Try the togas and tunics

For a spot of shopping look no further than the forum. This is like a town shopping centre where all sorts of things can be bought. Try out a white Roman toga or tunic for size and mix it up with the latest wool cloaks woven in yellows, blues and greens. Large bronze brooches and belt buckles are currently in fashion and favoured over the more delicate Roman jewellery in amber, emerald and gold.

See the slave stalls

The buying and selling of people will probably upset modern visitors, but the forum's slave stalls are a normal part of Londinium life. A slave's quality of life is entirely dependent on their owner. Slaves can end up being worked to death in a tin mine. Others can work as tutors for a wealthy family's children. Slaves can sometimes buy or be granted their freedom, but they can never become full Roman citizens. Their children, however, can.

Peek at the politicians

The seat of local government, the basilica, is a long, columned building running along one side of the forum. Here, officials meet in the round curia, or council chamber, and trials are held in the law court. The basilica does not yet hold guided tours, but make sure to poke your head around the corner for a peek inside.

The Public Baths

THE BUSINESS OF BATHING

After a sweaty day at the forum, why not visit one of Londinium's public baths to wash away the grime. The baths, however, are not only a place to get clean. You'll also do a little exercise, chat with the locals and have a bite to eat. By the end you'll be refreshed and ready to hit the streets again.

Prepare for the plunge

To bathe Roman style, you'll do some exercise before visiting a series of rooms with pools of different temperatures. First take a dip in the warm water of the tepidarium, before moving to the superhot caldarium and finishing off with a swim in the cold frigidarium. Now have a bite to eat at the bar and you're good to go.

Top Tip

Take your time

Londinium's baths are a social place where people go to catch up, hear all the latest gossip and do business deals. Therefore, set aside at least two hours for the baths and give yourself time to relax.

All are welcome

Public baths across the Roman Empire are kept cheap so everyone can visit. This means people from all walks of life mingle there: slaves, workers and aristocrats. However, men and women are not allowed to bathe together. Hours for women at Londinium's baths are between sunrise and noon; men between 2 pm and sunset.

TOP TIP

Forget your soap

It may seem strange, but no soap is used at the Londinium baths. Instead, bathers use the Roman method: rubbing oil into their skin and then, after working up a sweat, scraping off the excess oil and dirt. To do this, they use a curved iron tool called a strigil.

The Amphitheatre

A DAY AT THE GAMES

Visit the amphitheatre for an action-packed day of animal hunts, public executions and gladiator fights. Known as 'the games', this is fun-filled, family entertainment provided free of charge for all Londiniums. But be quick: there are only seats available for 7,000 spectators.

A bloody billing

The games follow a tried and tested day-long programme used at all amphitheatres across the Empire. Your day will start watching the fights between dangerous animals such as bears, lions and leopards. Then, other creatures will be hunted down by *venatores*, or animal hunters. At lunchtime, prisoners will be executed by crucifixion, or by being attacked by animals. The main attraction, however, is the gladiator fights that will take place in the afternoon. These are battles to the death between two gladiators who have been trained and armed to be evenly matched. Many gladiators achieve celebrity status and can win their freedom if they fight well.

Top Tip

Bring a barf bag

The games are enjoyed by people of all ages across the Roman Empire and are considered a normal part of everyday life. However, time-travellers beware – this is violence on a level that modern stomachs are not used to. It may pay to bring a barf bag!

Eyes on the emperor

The games are paid for by the emperor, governor or wealthy local judge as a gift to the people. This policy was introduced by Julius Caesar who believed he could win the affection of his people by offering them 'bread and circuses'. It is the job of the patron of the games to decide the fate of a wounded gladiator, by giving the thumbs up or down to his opponent. This is the signal that the gladiator should be spared, or finished off.

The Temple of Mithras

TRY A TEMPLE

Religion is an everyday feature of Londinium life and no stay would be complete without visiting a temple. But which temple will you choose? The Romans worship the official gods and goddesses, such as Jupiter and Mars, but also adopt the gods of the people they conquer. Londinium is a real mixing pot, with temples dedicated to Roman, British, Egyptian and Persian gods.

An underground cult

If you're feeling brave, try visiting a temple dedicated to one of the mystery cults, which are different from the main religions of the day. Originating in the east of the Empire, these cults carry out strange rituals to gods such as the Egyptian Isis and Persian Mithras. Admission to their dark, underground temples is usually only by special entry – but you may be able to sneak in for a few minutes at the end of a ceremony.

Top Tip

Boys only!

Sorry time-travelling girls, but the Temple of Mithras is for boys only. However don't be sad – Mithras followers have to promise to be pure, honest and behave at all times, which could get a little boring. Instead, why not visit the Temple of Isis? Isis is the Egyptian mother goddess believed to be able to grant eternal life, which doesn't sound too bad.

The business of religion

Romans take religion seriously as they believe that gods and goddesses control all areas of daily life. Offerings are made for favours, to say thanks, and to make amends for offending a god. The value of the gift depends on why it is being made, but common offerings include coins, jewellery, or the sacrifice of an animal. The Romans are happy for visitors to join in, regardless of their beliefs, so feel free to make an offering at the household shrine where you are staying.

The Fort

WALK TO THE FORT

It's hard to miss the massive fort in Londinium's north-west. Home to the Londinium legion, the fort is like a small town protected by thick, stone walls. A highly disciplined and well-drilled army is how Rome has been able to conquer and control its vast territories and create the largest empire in the world.

A popular model

The Londinium fort follows the standard rectangular Roman design with four gated entrances and two roads, which intersect the barracks, stables and administrative buildings inside. The fort is home to about a thousand soldiers, most of them legionaries. There is also a patch of ground for parades. This is an impressive sight, so ask the attending officer when the next parade will be.

Join the legion?

Men from all over the Empire join the Roman army (legion) for regular pay and a promised pension pot at the end. If you feel like giving up your comfortable, modern life, there may be room for you. However, before you sign up, think carefully. Roman legionaries have to serve for 25 years and are not allowed to get married during that time.

Watch out for the wall

Time-travellers to Londinium in the late CE 2nd century may notice a 6-m-high wall being constructed around the city. Designed to protect the city from foreign invasion, this wall is similar to one built by the Emperor Hadrian between England and Scotland.

Top Tip

Graffiti greetings

Graffiti is like the social media of the ancient world. If Londinium residents want to comment on the emperor, a gladiator, or their workmates, they scrawl it on to a city wall. One recent message by a bricklayer complaining about his colleague says: 'Austalis has been wandering off on his own every day for the last fortnight.'

VISIT QUICK!

Londinium residents consider their city to be a great symbol of Roman civilisation and culture. But there are bad omens on the horizon. Fortune tellers predict only one or two more centuries before the Roman Empire collapses and Londinium falls into disrepair. So time-travellers take note: visit now before your window of opportunity closes.

A Roman retreat

At the end of the 4th century, the Roman Empire was under attack. The Empire shrank as areas of land came under the control of invading tribes. Many of the soldiers who lived in Britannia were sent to defend Rome and some of the Roman citizens who lived in Londinium joined them. In 410, the city of Rome itself was overrun by Germanic tribes. The last remaining Roman soldiers left Londinium, leaving Britannia defenceless.

Goodbye advanced technologies

During its nearly 400-year existence, Londinium went from being a small town to the jewel of the Roman Empire in the north. However, after the Romans left, many of the technologies they introduced were lost. Knowledge of inside plumbing, underfloor heating, and the importance of drains was forgotten.

The new Londoners

During the 5th century, people gradually moved away from Londinium to the countryside. Britain came under increasing attack from people who lived in Denmark and north Germany. They travelled across the seas and up rivers to invade and settle parts of Britain. Londinium largely fell into ruins. It became important again during the reign of King Alfred the Great (ruled 849–899). It wasn't until the 13th century that London returned to the size of Roman Londinium.

GLOSSARY

Amphitheatre   A round, open building with raised seats surrounding a central space where shows were held.

Aristocracy   People in some countries who have a high social rank and special titles.

Cosmopolitan   A place that includes people from many different countries.

Counterfeit   An exact imitation of something valuable, such as coins.

Cult   A small religious group, especially one which is considered strange.

Ferment   A chemical process in food where the sugar turns to alcohol.

Fresco   A picture that is painted on a plastered wall when the plaster is still wet.

Gladiator   A man trained to fight with another gladiator for the entertainment of a crowd.

Governor   The official head of a country or region that is controlled by another country.

Legion   The Roman army.

Legionary   A Roman soldier.

Mutton   Meat that comes from an adult sheep.

Offering   Something precious given as part of a religious ritual, often in honour of a particular god.

Ointment   An oily substance rubbed onto the skin to heal it.

Roman Empire   The territories ruled by ancient Rome. At its height in the CE 1st century this included much of Europe, north Africa and south-west Asia.

Slave   Someone who is the property of another person. Prisoners of war or people captured in foreign countries were forced to become Roman slaves.

Thriving   To flourish or grow quickly and well.

Tribe   A social group made up of many families or clans that share the same culture, religion, beliefs and language.

FURTHER INFORMATION

Books

Britain in the Past: The Romans
Moira Butterfield, Franklin Watts, 2015

Found! Roman Britain
Moira Butterfield, Franklin Watts, 2017

The History Detective Investigates: London
Claudia Martin, Wayland, 2016

Invaders and Raiders: The Romans are coming!
Paul Mason, Franklin Watts, 2018

Websites

Try out Roman recipes, get Romans dressing up tips and watch a video interview with a Roman soldier on this interactive English Heritage website: **www.english-heritage.org.uk/members-area/kids/Roman-Britain-oma/**

Explore Londinium and see objects discovered from that time on the Museum of London website:
www.museumoflondon.org.uk/museum-london/permanent-galleries/roman-london

Find out about everyday life for Romans: **www.bbc.co.uk/schools/primaryhistory/romans/family_and_children/**

Read all about the vast Roman Empire:
www.historyforkids.net/ancient-rome.html

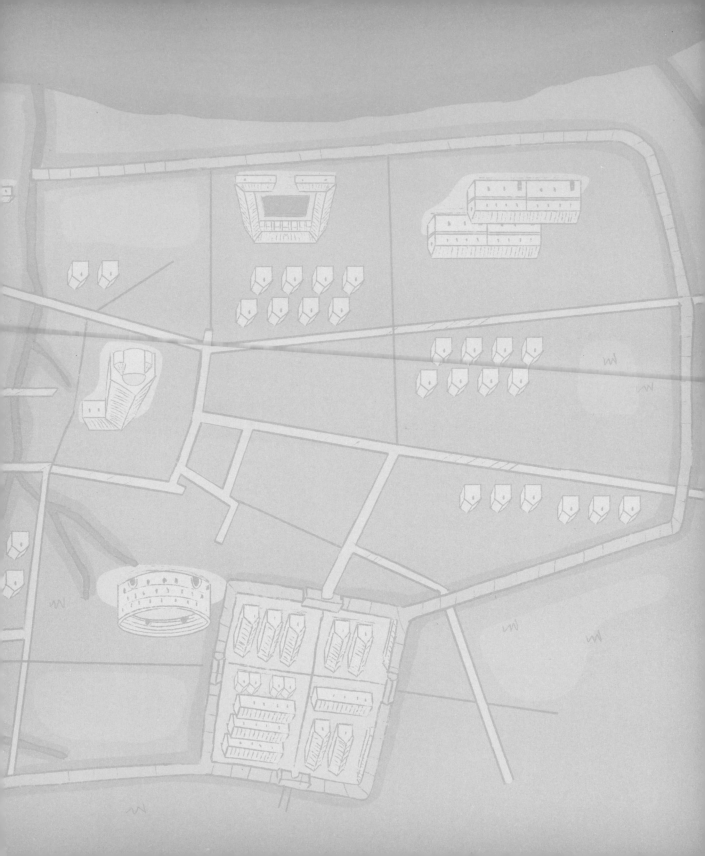